SOTHIS PRESS

Sothis Press aims to completely reimagine the public domain book with an emphasis on universal access to captivatingly designed print editions. We believe that technology will enable every single book ever published to be perpetually accessible in beautiful print form. Sothis is beauty in ubiquity.

www.sothis.press
1 Waterhouse Square, London, ECIN 2ST

A Key to the Exercises and Examples Contained in a Text-Book of Euclid's Elements

Books I-Vi, and XI

H. S. Hall

SOTHIS PRESS

ISBN 978-1-0365-1594-2

A Key to the Exercises and Examples Contained in a Text-Book of Euclid's Elements is in the public domain.

A KEY

TO THE

EXERCISES AND EXAMPLES

CONTAINED IN

A TEXT-BOOK OF EUCLID'S ELEMENTS

BOOKS I.—VI. AND XI.

A KEY

TO THE

EXERCISES AND EXAMPLES

CONTAINED IN

A TEXT-BOOK OF EUCLID'S ELEMENTS

BOOKS I.—VI. AND XI.

BY

H. S. HALL, M.A., AND F. H. STEVENS, M.A.

NEW EDITION

London

MACMILLAN AND CO., LIMITED

NEW YORK: THE MACMILLAN COMPANY

1905

First Edition 1892.
Reprinted 1899.
Second Edition 1901. Reprinted 1902, 1905.

GLASGOW: PRINTED AT THE UNIVERSITY PRESS
BY ROBERT MACLEHOSE AND CO. LTD.

PREFACE TO THE FIRST EDITION.

In preparing this Key two objects have been kept in view. It is intended first to save the time and lighten the work of teachers, and secondly to remove the difficulties of private students, leaving however sufficient demands upon their thought and intelligence to make the solutions in themselves a useful geometrical exercise. The Examples therefore have not in the majority of cases been worked out in detail, and the drawing of figures has been left to the reader.

The absence of figures may possibly give rise to some little difficulty in Examples which admit of a variety of cases, especially those in Book III. depending on angles in the same segment or on intersecting circles. It would of course be impossible within reasonable limits of space to deal separately with all the cases that may arise in every Example of this kind: we have therefore selected that case which we think would most naturally occur to a student in trying the problem for himself; and, when necessary, we have given some indication of the particular figure to which the proof refers. Other cases the student may easily, if he chooses, investigate for himself; the modifications which he will most frequently have to make will be the use of subtraction instead of addition of lines or angles, and the application of III. 22 instead of its kindred proposition III. 21.

H. S. HALL.
F. H. STEVENS.

October, 1892.

PREFATORY NOTE TO THE NEW EDITION.

THROUGHOUT this Key references have been altered so as to adapt it for use with the New Edition of our Euclid bearing the date 1900.

One other change has been made. A difficulty is often felt—especially in the case of private students—as to the form in which the solution of an elementary geometrical question should be presented; we have therefore given in fuller detail the solutions of the Exercises distributed through the text of Book I. The earlier of these now appear worked out at length.

H. S. H.
F. H. S.

August, 1900.

KEY TO EUCLID.

BOOK I.

EXERCISES.

Page 15.

1. Because A is the centre of the circle BCD,

$\therefore AF = AB.$ [Def. 15.]

And because B is the centre of the circle CAE,

$\therefore BF = AB.$ [Def. 15.]

Hence $AF = BF = AB$; [Ax. 1.]

so that the triangle AFB is equilateral.

2. In Prop. 1 it has been shewn that

AC and BC are each equal to AB;

and in Ex. 1 it has been shewn that

AF and BF are each equal to AB.

$\therefore$ AC, BC, AF, BF are equal.

Hence the fig. ACBF is a rhombus, since its angles are evidently not right angles. [Def. 32.]

3. With centre B, and radius BA, describe a circle, and produce AB to meet its circumference at C.

With centre A, and radius AC, describe a second circle PCQ; and from A draw any st. line to cut the circle PCQ at P.

Then AP shall be the required line.

Now AC is a diameter, and AB a radius, of the smaller circle;

$\therefore AC = \textit{twice}\ AB.$

And $AP = AC$, being radii of the circle PCQ; [Def. 15.]

$\therefore AP = \textit{twice}\ AB.$

4. Here OD = OE, being radii of the outer circle; [Def. 15.]
and OA = OB, being radii of the inner circle:
hence the remainder AD = the remainder BE. [Ax. 3.]

5. With centre P and radius PQ, describe a circle cutting AB (produced, if necessary) at X and Y.

Then since P is the centre of the circle,
∴ PQ = PX = PY; [Def. 15.]
that is, the distances of the *two* points X and Y from P are equal to PQ.

6. Here BD = AB, being sides of an equilateral triangle:
and AB = BC, by hypothesis;
∴ BD = BC, a radius of the circle CGH; [Ax. 1.]
∴ the point D is on the circumference.

7. On AC describe an equilateral triangle ADC.
With centre C, and radius CB, describe the circle BGH.
Produce DC to meet this circle at G.
With centre D, and radius DG, describe the circle GKF.
Produce DA to meet the circle GKF at F.
Then it may be shewn as in Prop. 2 that AF = BC.

8. From A and B draw AX and BY each equal to PQ. [I. 2.]
With centre A, and radius AX, describe a circle;
and with centre B, and radius BY, describe a second circle, cutting the first at C and D. Join AC, BC and AD, BD.

Then either ACB or ADB is the triangle required.
For AC = AX = PQ; [Def. 15 and Constr.]
and BC = BY = PQ.

Note that PQ must be greater than half AB, or the circles will not meet.

9. Draw AX and BY each double of the base AB; [Ex. 3.]
and from this point proceed as in the last example.

10. With centre A, and radius AN, describe a circle; and with centre B, and radius BM, describe a second circle, cutting the first at a point C.

Join AC, BC.

Then in the triangle ABC, AC = AN, and BC = BM. [Def. 15.]

Page 17.

1. In the $\triangle^s$ DAB, CBA,

Because { DA = CB, being sides of a square; [Def. 30.]
and AB is common;
and the $\angle$ DAB = the $\angle$ CBA, being right angles; [Def. 30.] }

$\therefore$ the $\triangle$ DAB = the $\triangle$ CBA in all respects; [I. 4.]

$\therefore$ DB = CA.

2. (i) In the $\triangle^s$ LBM, MCN,

Because { LB = MC, being halves of equal sides; [Ax. 7.]
and BM = CN, for the same reason;
and the $\angle$ LBM = the $\angle$ MCN, being right angles; [Def. 30.] }

$\therefore$ the $\triangle$ LBM = the $\triangle$ MCN in all respects; [I. 4.]

$\therefore$ LM = MN.

(ii) In the $\triangle^s$ ABM, DCM,

Because { AB = DC, being sides of a square; [Def. 30.]
and BM = CM, being halves of equal sides;
and the $\angle$ ABM = the $\angle$ DCM, being right angles; [Def. 30.] }

$\therefore$ the third side AM = the third side DM. [I. 4.]

(iii) In the $\triangle^s$ ABM, ADN,

Because { AB = AD, being sides of a square;
and BM = DN, being halves of equal sides;
and the $\angle$ ABM = the $\angle$ ADN, being right angles; }

$\therefore$ AM = AN. [I. 4.]

(iv) In the $\triangle^s$ BCN, DCM,

Because { BC = DC, being sides of a square;
and CN = CM, being halves of equal sides;
and the angle at C is common to both triangles; }

$\therefore$ the $\triangle$ BCN = the $\triangle$ DCM in all respects;

$\therefore$ BN = DM. [I. 4.]

3. In the △ˢ BAY, CAX,

Because { BA = CA, being sides of an isos. triangle;
and AY = AX, by hypothesis;
and the angle at A is common to both triangles;

∴ BY = CX. [I. 4.]

4. In the △ˢ BCD, ADC,

Because { BC = AD, by hypothesis;
and DC is common to both;
and the ∠ BCD = the ∠ ADC, by hypothesis;

∴ BD = AC. [I. 4.]

Page 21.

1. (i) Since AB = AD, being sides of a rhombus,

∴ the △ ABD is isosceles;

∴ the ∠ ABD = the ∠ ADB. [I. 5.]

(ii) Similarly it may be shewn that

the ∠ CBD = the ∠ CDB.

(iii) Hence, adding the equal angles in (i) and (ii),

the ∠ ABC = the ∠ ADC. [Ax. 2.]

2. In the △ ABC, since AB = AC, by hypothesis,

∴ the ∠ ABC = the ∠ ACB. [I. 5.]

And since DB = DC, [Hyp.]

∴ the ∠ DBC = the ∠ DCB. [I. 5.]

Hence the whole ∠ ABD = the whole ∠ ACD. [Ax. 2.]

3. As in the last example it may be shewn that

the ∠ ABC = the ∠ ACB;

and the ∠ DBC = the ∠ DCB.

∴ the remaining ∠ ABD = the remaining ∠ ACD. [Ax. 3.]

Page 25.

1. Let ABCD be a rhombus. Join BD.

Then in the △ˢ ADB, CDB,

Because { AD = CD, being sides of a rhombus; and BA = BC, for the same reason; and DB is common to both triangles;

∴ the ∠ DAB = the ∠ DCB. [I. 8.]

Similarly, by joining AC, it may be shewn that

the ∠ ADC = the ∠ ABC.

2. Join AC.

Then in the △ˢ ABC, CDA,

Because { AB = CD, by hypothesis, and BC = DA, ,, and AC is common to both;

∴ the ∠ ABC = the ∠ CDA. [I. 8.]

3. (i) Join AD.

Then in the △ˢ ABD, ACD,

Because { AB = AC, by hypothesis, and BD = CD, ,, and AD is common to both;

∴ the ∠ ABD = the ∠ ACD. [I. 8.]

(ii) The second case is similarly proved.

4. In the two △ˢ BAD, CAD,

Because { BA = CA, by hypothesis; and AD is common to both; also BD = CD;

∴ the ∠ BAD = the ∠ CAD, [I. 8.]

and the ∠ BDA = the ∠ CDA.

5. In Ex. 4 it has been shewn that the ∠ BAD = the ∠ CAD.

Then in the △ˢ BAE, CAE,

Because { BA = CA, [Hyp.] and AE is common to both; and the ∠ BAE = the ∠ CAE; [Proved.]

∴ the △ BAE = the △ CAE in all respects; [I. 4.]

so that BE = CE.

Page 27.

12. Let ABCD, EFGH be two squares in which AB = EF. Then the eight sides of the two squares must be all equal. Apply the sq. ABCD to the sq. EFGH, so that A falls on E, and AB falls along EF; then B must fall on F, for AB = EF. And since the ∠ ABC = the ∠ EFG, being right angles, ∴ BC must fall along FG; and, as before, C must fall on G. Similarly it may be shewn that D must fall on H. Hence the sq. ABCD coincides with the sq. EFGH, and is therefore equal to it in area.

13. Join BD, FH; and apply the fig. ABCD to the fig. EFGH, so that A falls on E, and AB falls along EF; then B must fall on F, for AB = EF. And since the ∠ BAD = the ∠ FEH [Hyp.], ∴ AD must fall along EH, and consequently D must fall on H, for AD = EH [Hyp.]. Then BD coincides with FH, and the △ DCB must therefore coincide with the △ HGF [I. 7], for DC = HG and BC = FG. Hence the fig. ABCD coincides with the fig. EFGH.

14. (i) In the △ˢ LBM, NCM,

Because { LB = NC, being halves of equal sides;
and BM = CM, for a similar reason;
also the ∠ LBM = the ∠ NCM; [I. 5.]

∴ LM = MN. [I. 4.]

(ii) In the △ˢ LBC, NCB,

Because { LB = NC, as above;
BC is common to both;
and the ∠ LBC = the ∠ NCB; [I. 5.]

∴ LC = NB. [I. 4.]

(iii) Join AM.

Then in the △ˢ ALM, ANM,

Because { AL = AN, being halves of equal sides;
LM = NM, proved in (i);
and AM is common to both;

∴ the ∠ ALM = the ∠ ANM. [I. 8.]

Page 28.

1. The point F might fall in any one of the following ways:

(i) above DE and below A, in which case the proof would still hold;

(ii) on the point A, in which case the construction fails;

(iii) above the point A, in which case FA *must be produced* in order to bisect the given angle.

2. In the △ˢ DFA, EFA,

Because { DF = EF, [Def. 24.]
and FA is common to both;
also DA = EA, by construction;

∴ the ∠ DFA = the ∠ EFA. [I. 8.]

3. Bisect the given angle by I. 9;
and by the same prop. bisect each of the half-angles.

Page 29.

1. Let ABC be the isosceles △, in which AB = AC;
and let AX bisect the vert. ∠ BAC, cutting BC at X.

Then in the △ˢ BAX, CAX,

Because { BA = CA, by hypothesis;
and AX is common;
and the ∠ BAX = the ∠ CAX, by hypothesis;

∴ BX = CX. [I. 4.]

2. Bisect the given straight line [I. 10]; and on the given base describe [as in Ex. 8, p. 15] an isosceles triangle, having each of its equal sides equal to the half-line.

Page 30.

1. Let X be any point in CF or CF produced. Join XD, XE.

Then in the △ˢ XCD, XCE,

Because { DC = CE, [Constr.]
and CX is common to both;
and the ∠ DCX = the ∠ ECX, being rt. angles;

∴ XD = XE. [I. 4.]

Page 32.

8. With centre P, and radius PQ, describe a circle; and with centre Q, and radius QP, describe a second circle intersecting the first at the points X and Y. Then it may be shewn, by the method of I. 1, that PXQY is the rhombus required.

9. With centre A, and radius equal to PQ, describe a circle; and continue the construction as in the last example. The problem is impossible unless PQ is greater than half AB; for otherwise the two circles will not intersect.

10. (i) and (iii) In the △s ODB, OEA,

Because { OD = OE, being radii of the outer circle;
OB = OA, " " inner circle;
and the ∠ at O is common to both;

∴ DB = EA; (i)
and the ∠ ODB = the ∠ OEA. (iii) [I. 4.]

(ii) Join AB.

Then since OA = OB [Def. 15], ∴ the △ OAB is isosceles;
and the equal sides OA, OB are produced to D and E:
∴ the ∠ BAD = the ∠ ABE. [I. 5, Part 2.]

Page 33.

1. Let A be the vertex, and BC the base, of an isosceles △ ABC; and let BC be bisected at D. Join AD.

Then in the △s BDA, CDA,

Because { BD = CD, [Hyp.]
and DA is common to both;
and AB = AC; [Hyp.]

∴ the ∠ ADB = the ∠ ADC; [I. 8.]
∴ these equal angles, being adjacent, are rt. angles.

2. Let ABC be an isosceles triangle, and X, Y the middle points of the equal sides AC, AB. Join BX, CY.

Then in the △s YBC, XCB,

Because { YB = XC, being halves of equal sides;
and BC is common to both;
also the ∠ YBC = the ∠ XCB; [I. 5.]

∴ YC = XB. [I. 4.]

3. In the base BC of the isosceles △ ABC, let D and E be two points such that BD = CE. Join AD, AE.

Then in the △s ABD, ACE,

Because { AB = AC, [Hyp.]
BD = CE, [Hyp.]
and the ∠ ABD = the ∠ ACE, [I. 5.]

∴ AD = AE. [I. 4.]

4. Let ABCD be a quad^l, in which AB = CD, and DA = BC. Join BD.

Then in the △s DAB, BCD,

Because { DA = BC, AB = CD, } by hypothesis,
and BD is common to both;

∴ the ∠ DAB = the ∠ BCD. [I. 8.]

Similarly, by joining AC, the ∠ ADC may be shewn equal to the ∠ ABC.

5. Join XY. Then in the △s XAY, XBY,

Because { XA = XB, AY = BY, } by hypothesis;
and XY is common to both;

∴ the ∠ XAY = the ∠ XBY; [I. 8.]
and the ∠ AXY = the ∠ BXY. [I. 8. Cor.]

6. Let ABCD be a rhombus, and BD a diagonal. Then, as in the last example, the △s ABD, CBD may be shewn equal in all respects, by I. 8.

∴ the ∠ ABD = the ∠ CBD; and the ∠ ADB = the ∠ CDB.

7. Let ABC be an isosceles triangle, and let BO, CO bisect the base ∠s ABC, ACB.

Then the ∠ ABC = the ∠ ACB, [I. 5.]
∴ the ∠ OBC = the ∠ OCB, being halves of equal angles;
∴ OB = OC. [I. 6.]

8. In the △s ABO, ACO,

Because { AB = AC, [Hyp.]
and BO = CO, by the last Ex.,
and the ∠ ABO = the ∠ ACO, being halves of equal angles;

∴ the ∠ BAO = the ∠ CAO. [I. 4.]

9. In the equilateral △ ABC, let D, E, F be the middle points of the sides BC, CA, AB, respectively. Join DE, EF, FD.

Then in the △s FAE, FBD,

Because {
FA = FB, [Hyp.]
and AE = BD, being halves of equal sides;
and the ∠ FAE = the ∠ FBD; [I. 5. Cor.]
}

∴ FE = FD. [I. 4.]

Similarly FD = DE; ∴ the △ FDE is equilateral.

10. Since AE = AF, and AB = AC; ∴ EB = FC.

Then in the △s EBC, FCB,

Because {
EB = FC,
and BC is common to both;
and the ∠ EBC = the ∠ FCB; [I. 5.]
}

∴ EC = FB. [I. 4.]

11. Let the diagonals of the rhombus ABCD intersect at X. Then, as in Ex. 6, the △s BAC, DAE may be shewn equal in all respects by I. 8. ∴ the ∠ BAC = the ∠ DAC.

Then in the △s BAX, DAX,

Because {
BA = DA, [Hyp.]
and AX is common;
and the ∠ BAX = the ∠ DAX;
}

∴ BX = DX; and the ∠ AXB = the ∠ AXD: [I. 4.]

and these angles, being adjacent, are rt. angles.

12. (i) In the △s ABY, ACX,

Because {
AB = AC, AY = AX, } by hypothesis;
and the angle at A is common;
}

∴ the ∠ ABY = the ∠ ACX. [I. 4.]

But the ∠ ABC = the ∠ ACB; [I. 5.]

∴ the remainder, the ∠ OBC, = the remainder, the ∠ OCB;

∴ OB = OC. [I. 6.]

(ii) Then the △s BAO, CAO may be shewn equal in all respects by I. 8; for BA = CA, BO = CO, and AO is common; ∴ the ∠ BAO = the ∠ CAO.

(iii) If AO, produced, meets BC at Z;
then the △ˢ BAZ, CAZ may be shewn equal in all respects by I. 4; for BA = CA, AZ is common; and the ∠ BAZ was proved equal to the ∠ CAZ.

∴ BZ = CZ; also the ∠ AZB = the ∠ AZC; and these, being adjacent, are rt. angles.

13. Let AB be the base, and P the length of the perp. Bisect AB at C; and at C draw CX at rt. angles to AB [I. 11], making CX equal in length to P [I. 3]. Join AX, BX. Then AXB will be the required triangle.

For in the △ˢ ACX, BCX,

Because { AC = BC, [Constr.]
and CX is common;
and the ∠ ACX = the ∠ BCX, being rt. ∠ˢ. }

∴ AX = BX.

14. Let A, B be the given points, and XY the given st. line.

Join AB, and bisect AB at C [I. 10].

At C draw CP at rt. angles to AB [I. 11], to meet XY at P. Join AP, BP.

Then the △ˢ ACP, BCP may be shewn equal in all respects, by I. 4, as in the last example;

∴ AB = BP.

The construction fails when AB is perp. to XY; for then CP will not meet XY.

Page 35.

1. In the △ ABC let BC be produced both ways to X and Y, and let the exterior ∠ˢ ABX, ACY be equal.

Then since the ∠ˢ ABC, ACB are respectively supplementary to the equal ∠ˢ ABX, ACY, [I. 18.]

∴ the ∠ ABC = the ∠ ACB;

∴ AB = AC. [I. 6.]

2. In the figure to this Example it is seen that the ∠ˢ XOB, BOY together make up half the ∠ˢ BOA, BOC, that is to say, half of two rt. angles [I. 13]. Hence the ∠ XOY is a rt. angle.

3. The ∠s AOX, COY, being respectively halves of the ∠s BOA, BOC, are together half of two rt. angles. [I. 13.]

Hence the ∠s AOX, COY are complementary.

4. The ∠s COX, XOA are supplementary; [I. 13.]

and the ∠ BOX = the ∠ XOA; [Hyp.]

∴ the ∠s BOX, COX are supplementary.

The second case is similarly proved.

Page 36.

1. In the △ CDA, because DA = DC, [Hyp.]

∴ the ∠ DAC = the ∠ DCA. [I. 5.]

Then in the △s DAO, DCO,

Because { DA = DC, [Hyp.]
and AO = CO, [Hyp.]
and the ∠ DAO = the ∠ DCO; [Proved.] }

∴ the ∠ DOA = the ∠ DOC; [I. 4.]

and these, being adjacent, are rt. angles.

Similarly the ∠s BOA, BOC are rt. angles.

Hence the two ∠s DOA, BOA, being together equal to two rt. angles, it follows that DO, OB are in one st. line. [I. 14.]

Page 39.

1. If any two lines BA, CA *meet at a point* A, and are cut by a third st. line BC, then the two interior ∠s ABC, ACB, on one side of BC are together less that two rt. angles.

2. In the △ ABC let BC be produced both ways to E and D.

Then the ∠s ABE, ABC together = 2 rt. angles; [I. 13.]

also the ∠s ACD, ACB together = 2 rt. angles;

hence the ∠s ABE, ABC, ACD, ACB = 4 rt. angles;

and of these, the ∠s ABC, ACB are together less than two rt. angles;

∴ the ∠s ABE, ACD are together greater than 2 rt. angles.

3. Join A to X, any point in BC.

Then the ext. ∠ AXC is greater than the int. opp. ∠ ABC; [I. 16.]
and the ext. ∠ AXB is greater than the int. opp. ∠ ACB.

∴ the ∠s ABC, ACB are together *less than* the ∠s AXC, AXB, which together = 2 rt. angles. [I. 13.]

Page 44.

1. From I. 17 it follows that every triangle must have *at least* two acute angles; ∴ the rt. angle is greatest angle; ∴ the hypotenuse, which is opposite to the rt. angle, is the greatest side. [I. 19.]

2. In the △ ABC, let the ∠ ABC = the ∠ ACB.

Then AB cannot be greater than AC, for then the ∠ ACB would be greater than the ∠ ABC. [I. 18.]

Similarly AB cannot be less than AC.

∴ AB = AC.

3. Since AB = AC, ∴ the ∠ ABC = the ∠ ACB. [I. 5.]

But in the △ ACD, the ext. ∠ ACB > the int. opp. ∠ ADB; [I. 16.]

∴ the ∠ ABD > the ∠ ADB;

∴ AD > AB. [I. 19.]

4. Let ABCD be a quadrilateral having AB the least side, and CD the greatest. Join BD.

Then since DA > AB [Hyp.], ∴ the ∠ ABD > ∠ ADB; [I. 18.]
and since DC > BC [Hyp.], ∴ the ∠ DBC > ∠ BDC;

∴ the whole ∠ ABC > the whole ∠ ADC.

So, by joining AC, the ∠ BAD may be shewn > the ∠ BCD.

5. Let AX be drawn to meet the base in X.

Then since AC is not greater than AB,

∴ AC is equal to, or less than AB;

∴ the ∠ ABC is equal to, or less than, the ∠ ACB. [I. 5 and 18.]

But the ext. ∠ AXB is greater than the int. opp. ∠ ACB, [I. 16.]

∴ the ∠ AXB is greater than the ∠ ABX;

∴ AB is greater than AX. [I. 19.]

6. Let AB be $>$ AC, then the $\angle$ ACB $>$ the $\angle$ ABC; [I. 18.]
hence the $\angle$ OCB $>$ the $\angle$ OBC; $\therefore$ OB $>$ OC. [I. 19.]

7. In the $\triangle$ ABC, BC is less than the sum of BA and AC. [I. 20.]
Take AC from each of these unequals; then the difference of BC and AC is less than BA.

8. In the quad[l] ABCD, let BA and CD meet at O.
Then in the $\triangle$ AOD, the sides AO, OD are together $>$ AD. [I. 20.]
To each of these unequals add AB, BC, CD:
then OB, BC, CO $>$ AD, AB, BC, CD.

9. Let ABC be the triangle, and O any point.
In the $\triangle$ OAB, OA + OB $>$ AB; [I. 20.]
in the $\triangle$ OBC, OB + OC $>$ BC; } add these results;
in the $\triangle$ OCA, OC + OA $>$ CA;
$\therefore$ *twice* the sum of OA, OB, OC $>$ the sum of AB, BC, CA,
i.e. the sum of OA, OB, OC $>$ half the perim. of the $\triangle$ ABC.

10. Let AC, BD be the diagonals of the the quad[l] ABCD.
Then from the $\triangle$ ABC, AB + BC $>$ AC; [I. 20.]
and from the $\triangle$ ADC, AD + DC $>$ AC;
$\therefore$, by addition, the perimeter $>$ *twice* AC.
Similarly, the perimeter $>$ *twice* BD.
$\therefore$, by addition, *twice* perimeter $>$ *twice* sum of diagonals.
$\therefore$ perimeter $>$ sum of diagonals.

11. In the $\triangle$ ABC, let AX bisect the $\angle$ BAC and meet BC at X.
Then the ext. $\angle$ AXC $>$ the int. opp. $\angle$ BAX; [I. 16.]
$\therefore$ the $\angle$ AXC is also $>$ the $\angle$ CAX; [Hyp.]
$\therefore$ AC $>$ XC. [I. 19.]
Similarly, AB $>$ BX;
$\therefore$, by addition, AB, AC together $>$ BC.

12. Produce AD, towards the base, to any point X.
Then the ext. $\angle$ BDX $>$ the int. opp. $\angle$ BAD; [I. 16.]
and the ext. $\angle$ CDX $>$ the int. opp. $\angle$ CAD;
$\therefore$, by addition, the whole $\angle$ BDC $>$ the whole $\angle$ BAC.

13. Let O be any point within the △ ABC.

$$\left.\begin{aligned}\text{Then } AO + BO &< AC + BC,\\ \text{and } BO + CO &< BA + CA,\\ \text{and } CO + AO &< CB + AB,\end{aligned}\right\}$$ [I. 21.]

∴, by addition, *twice* the sum of OA, OB, OC is less than *twice* the sum of AB, BC, CA.

$$\therefore\ OA + OB + OC < AB + BC + CA.$$

Page 47.

1. See fig. to Prop. 22. Let FG be the given base, and A, C the given st. lines to which the remaining sides of the triangle are to be equal.

Produce FG both ways, and cut off FD equal to A [I. 3], and GH equal to C.

With centre F, and radius FD, describe the circle DLK;

With centre G, and radius GH, describe the circle MHK, cutting the first circle at K. Join FK, GK. Then KFG is the triangle required. [Proof, as in Prop. 22.]

If A is greater than the sum of FG and C, then FK is greater than FH; and the second circle will fall *within* the first, so that there would be no point of intersection K. Similarly other cases may be examined.

Page 50.

1. In the two △s BXA, CXA, let AB be greater than AC.

Then since BX = CX, and AX is common;

and AB > AC, ∴ the ∠ AXB > the ∠ AXC: [I. 25.]

∴ the ∠ AXB is more than half the sum of the ∠s AXB, AXC, that is, more than half of two rt. angles. [I. 13.]

∴ the ∠ AXB is obtuse.

Similarly it may be shewn that if AB < AC, the ∠ AXB is acute.

Page 55.

1. Since AB = AC, ∴ the ∠ ABC = the ∠ ACB. [I. 5.]

Then in the $\triangle^s$ YBC, XCB,

Because { the ∠ YBC = the ∠ XCB, [Proved.]
and the ∠ YCB = the ∠ XBC, being half of equal $\angle^s$;
and BC is common to both;

∴ the △ YBC = the △ XCB in all respects. [I. 26.]

2. In the isosceles △ ABC, let BX, CY be drawn perp. respectively to the equal sides AC, AB.

Then in the $\triangle^s$ YBC, XCB,

Because { the ∠ YBC = the ∠ XCB, [I. 5.]
and the ∠ BYC = the ∠ CXB, being rt. angles;
and BC is common to both;

∴ CY = BX. [I. 26.]

3. Let O be any point on the bisector of the ∠ BAC.

Draw OP, OQ perp. respectively to AB, AC.

Then in the $\triangle^s$ OAP, OAQ,

Because { the ∠ OAP = the ∠ OAQ, [Hyp.]
and the ∠ OPA = the ∠ OQA, being rt. angles,
and OA is common to both;

∴ OP = OQ. [I. 26.]

4. In the $\triangle^s$ AOX, BOY,

Because { the ∠ AOX = the ∠ BOY, being vert. opp. $\angle^s$, [I. 15.]
and the ∠ AXO = the ∠ BYO, being rt. angles,
and AO = BO; [Hyp.]

∴ AX = BY. [I. 26.]

5. In the △ ABC, let AX, the bisector of the ∠ BAC, be perp. to the base BC.

Then in the $\triangle^s$ AXB, AXC,

Because { the ∠ BAX = the ∠ CAX, [Hyp.]
and the ∠ AXB = the ∠ AXC, being rt. $\angle^s$; [Hyp.]
and AX is common to both;

∴ AB = AC. [I. 26.]

6. Let AB be the given st. line, and P the given point.

Draw PC perp. to AB, and draw PD, PE to meet AB on the same side of PC; PD being the nearer line to PC.

Then in the △ PCD, the ∠ PCD is a rt. angle, [Constr.]

∴ the ∠ PDC is acute [I. 17]; ∴ PC < PD. [I. 19.]

And since the ∠ PDC is acute, ∴ the ∠ PDE is obtuse; [I. 13.]

∴ the ∠ PDE > the ∠ PED; ∴ PE > PD.

If PF be drawn on the other side of PC, making the ∠ CPF equal to the ∠ CPD, then the △ˢ CPD, CPF may be shewn to be identically equal [I. 26]; so that PF = PD. And, as before, it may be shewn that PF is greater than any line nearer PC, and less than any line more remote.

7. In the △ˢ PKH, P′KH,

Because { PH = P′H, [Constr.]
HK is common to both;
and the ∠ PHK = the ∠ P′HK, being rt. angles;

∴ the ∠ PKH = the ∠ P′KH [I. 4.]
= the vert. opp. ∠ AKQ. [I. 15.]

8. Let AB be the first given st. line, and OC, OD the given intersecting lines.

Bisect the ∠ COD by a line which meets AB in X.

From X draw XP, XQ perp., respectively to OC, OD.

Then the △ˢ OXP, OXQ may be shewn to be equal in all respects [I. 26]; ∴ XP = XQ.

If AB is parallel to the bisector of the ∠ COD, the point X cannot be found.

9. Let O be the first given point, and A, B the others.

Join AB, and bisect AB at C [I. 10]. Through O and C draw a st. line OCD; and draw AP, BQ perp. to OCD. [I. 12.]

Then the △ˢ ACP, BCQ may be shewn equal in all respects [I. 15, and 26]. ∴ AP = BQ.

The construction is impossible if O, A, B, are in the same st. line.

Page 60.

1. In the $\triangle^s$ AOC, BOD,

$\therefore \left\{ \begin{array}{l} \left. \begin{array}{r} AO = BO, \\ \text{and } OC = OD, \end{array} \right\} \text{by hypothesis;} \\ \text{and the } \angle\, AOC = \text{the } \angle\, BOD, \text{ being vert. opp. } \angle^s; \end{array} \right.$ [I. 15.]

$\therefore$ the $\angle$ OAC = the $\angle$ OBD ; [I. 4.]

and these are alternate angles; $\therefore$ AC and BD are parallel. [I. 27.]

2. Let each of the st. lines AB, CD be perp. to a st. line which meets them at P and Q.

Then the $\angle^s$ APQ, PQD are equal, being rt. angles; and since these angles are alternate, $\therefore$ AB and CD are parallel. [I. 27.]

3. Let PQ meet the parallels AB, CD at P and Q, and let PQ be perp. to AB.

Then since AB and CD are par[l], and PQ meets them,

$\therefore$ the $\angle$ PQD = the alt. $\angle$ APQ [I. 29.]
= a rt. angle. [Hyp.]

4. Let AB, CD, which intersect at E, be respectively par[l]. to PQ, RS, which intersect at O, the letters being so taken that the $\angle^s$ CEB, ROQ are both acute or both obtuse. Let CE, produced if necessary, meet PQ at X.

Then since AB, PQ are par[l]., and CEX meets them,

$\therefore$ the $\angle$ CEB = the $\angle$ CXQ ; [I. 29.]

and since CX, RS are par[l]., and PXQ meets them,

$\therefore$ the $\angle$ ROQ = the $\angle$ CXQ ;

$\therefore$ the $\angle$ CEB = the $\angle$ ROQ.

Similarly the $\angle$ CEA = the $\angle$ ROP.

Page 63.

1. Let XY cut the equal sides AB, AC of the isosceles $\triangle$ ABC at X and Y.

Then since XY, BC are par[l]., and AXB meets them,

$\therefore$ the ext. $\angle$ AXY = the int. opp. $\angle$ ABC. [I. 29.]

Similarly the $\angle$ AYX = the $\angle$ ACB.

But the $\angle$ ABC = the $\angle$ ACB [I. 5]; $\therefore$ the $\angle$ AXY = the $\angle$ AYX.

2. Let P be any point in the bisector of the ∠ AOB; and let PQ be drawn par[l]. to OB.

Then since OB, PQ are par[l]., and OP meets them,

∴ the ∠ QPO = the alt. ∠ POB [I. 29.]
= the ∠ POQ, [Hyp.]
∴ QO = QP. [I. 6.]

3. Let O be the given point, AB the given st. line, and K the given ∠.

At P, any point in AB, make the ∠ BPQ equal to K. [I. 23.]
Through O draw OX par[l] to AB. [I. 31.]
Then OX will make the required angle with AB. [Proof by I. 29.]

4. Draw AD perp. to BC, and from the △[s] BAD, CAD, shew, by I. 26, that the ∠ BAD = the ∠ CAD.

Then AD and ZYX, being both perp. to BC, are parallel;

∴ the ∠ BAD = the alt. ∠ AYZ;
also the ∠ CAD = the int. opp. ∠ AZY; } [I. 29.]
∴ the ∠ AYZ = the ∠ AZY; ∴ AY = AZ. [I. 6.]

5. In the △ ABC, let BA be produced to D; and let AX bisect the ∠ CAD, and be par[l] to BC.

Then the ext. ∠ XAD = the int. opp. ∠ CBA;
and the ∠ XAC = the alt. ∠ BCA; } [Hyp. and I. 29.]
but the ∠ XAD = the ∠ XAC [Hyp.]; ∴ the ∠ CBA = the ∠ BCA;
∴ AB = AC. [I. 6.]

Page 65.

1. (i) Let AD be drawn through the vertex A par[l] to base BC.
Then ∠ BCA = alt. ∠ CAD. [I. 29.]
∴ the three ∠[s] of △ ABC = ∠[s] CBA, BAD
= 2 rt. ∠[s]. [I. 29.]

(ii) Let AX be drawn to a pt. X in the base BC.
Then ext. ∠ AXC = sum of int. ∠[s] XBA, BAX; [I. 32.]
and ext. ∠ AXB = sum of int. ∠[s] XAC, ACX.
∴, by addition, the three ∠[s] of △ ABC = sum of ∠[s] AXB, AXC
= 2 rt. ∠[s]. [I. 13.]

2. Let ABC be the △, and let the base BC be produced to X and Y.

Then ext. ∠ XBA = sum of ∠s BAC, BCA; [I. 32.]
and ext. ∠ YCA = sum of ∠s CBA, BAC.

∴ the sum of the two ext. ∠s = the three angles of △ ABC, together with ∠ BAC,
= 2 rt. ∠s + ∠ BAC. [I. 32.]

3. Let the lines XP, XQ be respectively perp. to the lines AP, AQ, which form an acute angle at A. Let X be taken outside the ∠ PAQ, and let PX meet AQ in O.

Then ∠ QXO = compt. of ∠ QOX
= compt. of ∠ POA
= ∠ PAO.

4. Let △ ABC be rt. angled at C, and let D be the middle pt. of the hypotenuse AB.

Draw DE, DF perp. to AC, BC, and therefore parl to BC and AC respectively.

Then in △s AED, DFB,

Because { ∠ ADE = int. ∠ DBF,
∠ AED = ∠ DFB, being rt. ∠s,
and AD = DB; [Hyp.]

∴ DF = AE. [I. 26.]

Also from △s EDC, FDC it may be shewn that DF = EC. [I. 26.]

∴ AE = EC.

Then in △s DEA, DEC,

Because { AE = EC,
DE is common to both,
and ∠ AED = ∠ CED, being rt. ∠s;

∴ DA = DC. [I. 4.]

But DA = DB [Hyp.]; ∴ DC = DB.

Thus each of the △s ADC, BDC is isosceles, and DC is half of AB.

6. Let ABC be the rt. ∠. On BC describe an equilateral △ BDC; bisect ∠ DBC by BE; then ∠ ABC is trisected by BD, BE.

For ∠ DBC = two-thirds of a rt. ∠; [I. 32.]

∴ ∠ˢ DBE, EBC are each one-third of a rt. ∠.

7. Let BO, CO be the bisectors of the base angles of an isosceles △ ABC.

Then ∠ BOC = suppᵗ. of sum of ∠ˢ OBC, OCB [I. 32.]
= suppᵗ. of ∠ ABC [Hyp.]
= exterior base angle.

8. Let ABCD be the quadril. with ∠ˢ at D and C bisected by DO, CO.

Then twice the sum of ∠ˢ DOC, ODC, OCD = 4 rt. ∠ˢ;
that is, twice ∠ DOC together with ∠ˢ BCD, CDA = 4 rt. ∠ˢ.
But the four angles of the quadril. = 4 rt. ∠ˢ;
∴ twice ∠ DOC = ∠ˢ CBA, BAD.

Page 67.

1. (i) The six ∠ˢ, with 4 rt. ∠ˢ, are together equal to 12 rt. ∠ˢ, [I. 32, Cor. 1.]
that is, the six ∠ˢ together = 8 rt. ∠ˢ;
∴ each ∠ = $\frac{4}{3}$ rt. ∠.

(ii) The eight ∠ˢ, with 4 rt. ∠ˢ, are together equal to 16 rt. ∠ˢ.
∴ each ∠ = $\frac{3}{2}$ rt. ∠.

2. The *interior* ∠ = $\frac{4}{3}$ rt. ∠;
∴ the exterior ∠ = 2 rt. ∠ˢ − $\frac{4}{3}$ rt. ∠ = $\frac{2}{3}$ rt. ∠.

3. Let ABCDE be the rectil. fig. Join DA, DB; then the five-sided fig. is divided into 3 △ˢ. Thus the interior angles are together equal to 6 rt. ∠ˢ. ∴ the int. ∠ˢ together with 4 rt. ∠ˢ = 10 rt. ∠ˢ. Similarly the corollary may be proved for a fig. of any number of sides.

4. The n angles + 4 rt. ∠ˢ = $2n$ rt. ∠ˢ.
∴ the n angles = $(2n - 4)$ rt. ∠ˢ,
∴ each ∠ = $\frac{2n-4}{n}$ right angles.

5. Take the fig. of page 66. Let AB, DC meet in G; BC, ED in H; CD, AE in K; DE, BA in L; EA, CB in M.

Then, by Prop. 32, Cor. 2 the base ∠ˢ of the exterior △ˢ are together equal to 8 rt. ∠ˢ.

And the sum of *all* the ∠ˢ of these △ˢ = twice as many rt. ∠ˢ as the fig. has sides.

∴ the ∠ˢ at the vertices together with 8 rt. ∠ˢ = twice as many rt. ∠ˢ as the fig. has sides.

Page 70.

1. Let ABCD be a par^{m}, having the ∠ at A a rt. ∠.

Then the sum of ∠ˢ at A and D = 2 rt. ∠ˢ, [I. 29.]

∴ ∠ at D = rt. ∠.

Similarly the other two angles may each be proved to be rt. ∠ˢ.

2. In quadril. ABCD, if BD is joined,

Because { AB = CD, BC = AD, BD is common,

∴ △ˢ BCD, BAD are identically equal. [I. 8, Cor.]

∴ ∠ ABD = ∠ BDC,

∴ AB is par^{l}. to DC. [I. 27.]

∴ AB and DC are equal and parallel; hence, by Prop. 33, it follows that ABCD is a par^{m}.

3. In the quadril. ABCD,

let ∠ DAB = ∠ BCD,

and ∠ ABC = ∠ ADC;

then the sum of two adjacent ∠ˢ = ½ sum of the ∠ˢ of fig. ABCD

= 2 rt. ∠ˢ; [I. 32.]

∴ the opposite sides are parallel. [I. 28.]

4. By Ex. 2 the fig is a par^{m}. *with one right* ∠;

∴, by Ex. 1, all its angles are rt. ∠ˢ.

[For Examples 5-10 inclusive use diagram on p. 69.]

5. Let AD meet BC in O, then in $\triangle^s$ AOB, COD,

Because $\begin{cases} \angle \text{ABO} = \text{alt. } \angle \text{OCD}, \\ \angle \text{AOB} = \text{vert. opp. } \angle \text{COD}, \\ \text{and AB} = \text{CD}, \end{cases}$ [I. 15.]

$\therefore$ $\triangle^s$ AOB, COD are identically equal, [I. 26.]

$\therefore$ AO = OD, and CO = OB.

6. $\triangle^s$ AOB, COD are equal in all respects [I. 15 and I. 4]. Hence AB, CD are equal and may be shewn parallel. Hence the fig. ABCD is a par^{m}. [I. 33.]

7. $\angle$ ABC = $\angle$ ACB, being halves of equal $\angle^s$. $\therefore$ AB = AC.

8. In $\triangle^s$ ABD, BDC,

Because $\begin{cases} \text{AB} = \text{DC}, \\ \text{AD} = \text{BC}, \\ \text{and BD is common}, \end{cases}$

$\therefore$ $\angle$ ABD = $\angle$ BDC [I. 8]; $\therefore$ each is a rt. $\angle$. [I. 29.]

9. In $\triangle^s$ ABD, BDC, if $\angle^s$ ABD, BDC are not equal, AD is not equal to BC [I. 25].

10. Let the line meet AB in P, CD in Q; then $\triangle^s$ OPB, OQC are equal in all respects [I. 26], $\therefore$ OP = OQ.

11. Let ABCD, EFGH be par^{ms}. in which

AB = EF,
BC = FG;
and $\angle$ ABC = $\angle$ EFG.

Then $\triangle$ ABC may be made to coincide with $\triangle$ EFG [I. 4], and $\triangle$ ADC will coincide with $\triangle$ EHG. [I. 7.]

12. This follows at once by I. 4.

13. Let AP, CQ be perp. to diag. BD.

Then $\angle$ ADP = alt. $\angle$ QBC,
$\angle$ APD = $\angle$ CQB, being rt. $\angle^s$,
and AD = BC;

$\therefore$ $\triangle^s$ APD, BQC are identically equal, [I. 26.]

$\therefore$ AP = CQ.

14. Since AX is equal and par^{l}. to YC, it follows that AYCX is a par^{m}. [I. 33.]

Page 71.

1. The construction consists of *three* steps: (i) joining A to an extremity of BC, (ii) describing an equilat. △ on the joining line, (iii) producing two sides of the equilat. △.

Now each of these steps may be performed in *two* ways: for (1) A might be joined to *either* extremity of BC, (2) the equilat. △ might be described on *either* side of the joining line, and (3) the two sides might be produced in *either* direction. Hence the number of constructions is 2 × 2 × 2, or 8.

The exceptional case is when A is situated at the vertex of an equilat. △ on BC.

2. In fig. to Prop. 15 let EX, EY bisect ∠ˢ BEC, AED respectively.

Then ∠ˢ XEC, CEA, AEY = ∠ˢ XEB, BED, DEY respectively.

Thus sum of ∠ˢ XEC, CEA, AEY = 2 rt. ∠ˢ;

∴ EX, EY are in same st. line.

3. From equal △ˢ BAE, CEF,

AB = CF,

∠ BAE = alt. ∠ ECF;

∴ AB is parˡ. to CF. [I. 27.]

∴ AF and BC are equal and parˡ. [I. 33.]

Thus ABCF is a parᵐ. and △ˢ ABC, AFC are equal in all respects. [I. 34.]

4. ∠ ACD is greater than int. opp. ∠ ABC, [I. 16.]

that is, ∠ ACD is obtuse.

5. See solution of Ex. 4, p. 67.

6. In fig. to Prop. 21, suppose BD, CD bisect base ∠ˢ, and let AD be produced to F;

then ∠ BDF = sum of ∠ˢ DBA, BAD;

and ∠ CDF = sum of ∠ˢ DCA, CAD.

Hence ∠ BDC = the angle at A together with half the sum of the base angles.

7. In △ ABC let the external bisectors of ∠^{s} B and C meet at F, and the internal bisectors at E.

Then by Ex. 2, page 35 each of ∠^{s} EBF, ECF is a rt. ∠.

∴ ∠ F is the supp^t. of ∠ E;

that is, ∠ F = sum of ∠^{s} EBC, ECB.

8. Let the st. lines intersect at O, and let P be the pt. from which perps. PX, PY are drawn.

If PX and PY are equal it may be shewn [I. 26] that OP bisects both the ∠^{s} XOY, XPY.

But if PX is greater than PY, let the bisector of the ∠ XOY meet PX in Q; and let the bisector of the the ∠ YPQ meet OY at S. Draw QR perp. to OY and therefore par^l. to PY.

Then the △^{s} OQR, OQX are equal in all respects. [I. 26.]

∴ ∠ RQO = ∠ XQO.

But since RQ is par^l. to YP;

∴ ∠ RQX = ∠ YPX;

and the halves of these ∠^{s} are equal; that is, ∠ OQX = ∠ SPX.

∴ SP is par^l. to OQ. [I. 28.]

9. By hypothesis, AP = AQ;

∴ ∠ APQ = ∠ AQP;

∴ each is half the supp^t. of ∠ BAC.

Hence ∠ APQ = ∠ ACB.

∴ PQ is par^l. to BC. [I 28.]

10. Join AD, BE, CF.

Then because AB is equal and par^l. to DE,

∴ AD is equal and par^l. to BE. [I. 33.]

But BC is equal and par^l. to EF;

∴ BE is equal and par^l. to CF;

that is, AD is equal and par^l. to CF;

∴ AC is equal and par^l. to DF. [I. 33.]

11. This will be easily seen from the adjoining diagram.

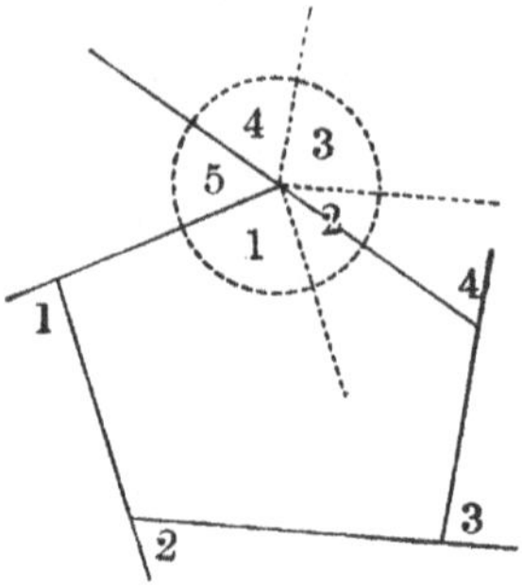

12. In the quadril. ABCD, let AB be par[l]. to CD, and AD equal to BC. Draw BE par[l]. to AD;

then ABED is a par[m]. and BE = AD = BC.

∴ ∠ BCD = ∠ BED = the supp[t]. of ∠ DAB.

Again, since the four ∠[s] of the fig. are together equal to four rt. ∠[s], the other pair of opp. ∠[s] are supplementary.

Join AC, BD.

Then in △[s] DEB, ABC,

∠ DEB = supp[t]. of ∠ BCD

= ∠ ABC; [I. 29.]

and DE, EB = AB, BC respectively;

∴ DB = AC. [I. 4.]

Page 79.

1. (i) The △[s] ABD, ACD are equal in area. [I. 37.]
Take away △ AKD from each and the remainders are equal.

(ii) Par[m]. EBCA = par[m]. DBCF.
Take away △ BKC from each and the remainders are equal.

2. See solution to Ex. 3 on p. 71.

3. Let AB be the base of given △ ABC, and let PQ be the st. line in which the vertex is to lie. Through C draw CD par[l] to AB meeting PQ in D; then ADB is the required △. [I. 37.]

4. Let AB be the base of the given △ ABC. Through C draw CD par[l]. to AB. Bisect AB in E, and through E draw EF perp. to AB meeting CD in F. Join FA, FB.

Then △ AFB = △ ABC. [I. 37.]

Again, in △[s] AEF, BEF,

Because { AE = BE, EF is common, and ∠ AEF = ∠ BEF, being rt. ∠[s];

∴ AF = BF. [I. 4.]

5. The △[s] are on equal bases and of equal altitude.

6. In the fig. to I. 34 let the diagonals intersect in E. Then, by Ex. 5 on p. 70, E is the middle pt. of each diagonal.

∴ △ AEC = △ AEB = △ BED = △ DEC.

7. The △ ABX = △ ACX; and △ YBX = △ YXC. [I. 38.]

Hence △ ABY = △ ACY.

8. Join BD cutting AC in E, then AE is a median of △ ABD. Thus the result follows by Ex. 7.

9. Since the equal sides contain supplementary angles the two △[s] can be placed having one side in common and the other two equal sides in same st. line. Thus we have two △[s] of same altitude on equal bases.

10. Let △ ABC have sides AB, BC bisected in X, Y respectively;

then △ BXY = △ AXY, and △ BXY = △ CXY; [I. 38.]

∴ AXY = △ CXY;

∴ XY is par[l]. to BC. [I. 39.]

11. Join AD, BC. Then by adding △ AOD to the equal △[s] AOC, DOB,

△ DAC = △ DAB;

∴ BC is par[l]. to AD. [I. 39.]

12. △ AEF = △ ABC [I. 38.]

= △ DEF; [Hyp.]

∴ AD is par[l]. to EF. [I. 39.]

Page 80.

1. Par^{m}. AY = half par^{m}. AC,
and △ AZB = half par^{m}. AY ; [I. 41.]
∴ △ AZB = one-fourth par^{m}. AC.

2. Let BD be a diagonal of the given sq. ABCD. At B draw BE perp. to DB meeting DC produced in E. Then in △^{s} DCB, ECB,

Because { ∠ DBC = ∠ CBE, each being half a rt. ∠,
∠ DCB = ∠ ECB, being rt. ∠^{s},
and BC common ;

∴ △^{s} DCB, ECB are identically equal. [I. 26.]
∴ DB = BE.
Also △ DBE = twice △ DCB = given square.

3. △^{s} AXB, BYC are each of them half of the given par^{m}.

4. Through P draw XY par^{l}. to AB or DC.
Then △ APB is half par^{m}. AY,
and △ DPC is half par^{m}. XC ;
that is, sum of △^{s} APB, DPC is half par^{m}. AC.

Page 81.

1. Let ABCD be the given square. Join BD. Through C draw CE par^{l}. to BD meeting AB produced in E ;
then par^{m}. DBCE = sq. AC. [I. 35.]
Also ∠ BDC = half a rt. ∠.

2. Let ABCD be the given par^{m}.
With centre D and radius DC describe a circle cutting AB in E.
Draw CF par^{l}. to DE meeting AB produced in F.
Then EDCF is a rhombus and it is equal to par^{m}. ABCD.
If DC is less than perp. from D to AB the circle will not meet AB and the construction fails.

Page 83.

(i) By adding par^{m}. EH to each of equal par^{ms}. KD, BK, par^{m}. ED = par^{m}. BH.

(ii) △^{s} AKD, AKB are respectively halves of the equal par^{ms}. ED, BH.

Page 88.

See solution of Ex. 29, p. 120.

Page 92.

1. (i) ∠ GBA = alt. ∠ AHC, each being half a rt. ∠,

∴ BG is par^{l}. CH. [I. 27.]

(ii) ∠ FAB = half a rt. ∠

= ∠ KAC ;

∴ ∠^{s} FAB, BAC, CAK together = 2 rt. ∠^{s},

∴ FA, AK are in one st. line.

(iii) Let FC meet AB in M and AD in N.

Then from equal △^{s} FBC, ABD,

∠ BFC = ∠ BAD ;

also ∠ FMB = ∠ AMN ; [I. 15.]

∴ ∠ FBA = ∠ ANB, [I. 32.]

∴ AD is at rt. ∠^{s} to FC.

(iv) △ GAH is obviously equal to △ ABC in all respects by I. 4.

Again, since ∠^{s} FBA, DBC are rt. ∠^{s},

∴ ∠ FBD is supp^{t}. of ∠ ABC,

Similarly, ∠ KCE is supp^{t}. of ∠ ACB.

Hence the result follows by Ex. 9, p. 79.

2. Take the case of the exterior squares ;

then in △^{s} CAG, BAH,

Because { CA = HA,
AG = AB,
and ∠ CAG = BAH ;

∴ CG = BH. [I. 4.]

The other case is similar.

3. It is easy to see that △^{s} ACX, BCY are equal in all respects. [I. 4.]

∴ AX = BY. Similarly BY = CZ.

4. Let BD be the diagonal of given sq. ABCD,

then sq. on DB = sum of sqq. on DC, BC

= twice sq. on DC.

5. In $\triangle^s$ ABX, ACX,

Because $\begin{cases} \angle \text{ABX} = \angle \text{ACX}, \\ \angle \text{AXB} = \angle \text{AXC, being rt. } \angle^s. \\ \text{and AX is common;} \end{cases}$

∴ $\triangle^s$ ABX, ACX are equal in all respects: [I. 26.]

∴ BX = XC; or BC is twice BX.

∴ sq. on BC = 4 times sq. on BX,

that is, sq. on AD = 4 times sq. on BX.

But sq. on AB = sum of sqq. on BX, AX;

∴ sq. on AX = 3 times sq. on BX.

6. Let AB, CD be sides of the two squares; at B draw BE perp. to AB and equal to CD. Join AE.

Then sq. on AE = sum of sqq. on AB, BE

= sum of sqq. on AB, CD.

7. Sq. on AB = sum of sqq. on AX, BX,

sq. on AC = sum of sqq. on AX, XC;

∴ diff. of sqq. on AB, AC = diff. of sqq. on BX, XC.

8. By Ex. 7,

sq. on BZ − sq. on AZ = sq. on OB − sq. on OA,

sq. on AY − sq. on CY = sq. on OA − sq. on OC,

sq. on CX − sq. on BX = sq. on OC − sq. on OB;

∴, by addition,

sum of sqq. on BZ, AY, CX − sum of sqq. on AZ, CY, BX = 0,

or, sum of sqq. on BZ, AY, CX = sum of sqq. on AZ, CY, BX.

9. Sq. on BQ = sum of sqq. on AB, AQ,

sq. on PC = sum of sqq. on AC, AP;

∴ sum of sqq. on BQ, PC = sum of sqq. on AB, AC, together with sum of sq. on AQ, AP.

= sum of sqq. on BC, PQ.

10. Let the medians be BQ, CP; then it may be shewn that BC is double of PQ. [See solution of Ex. 3, p. 105.]

∴ sq. on BC = 4 times sq. on PQ.

And by Ex. 9,

sum of sqq. on BQ, CP = sum of sqq. on BC, PQ
= 5 times sq. on PQ.

THEOREMS AND EXAMPLES ON BOOK I.

I. On the Identical Equality of Triangles.

Page 98.

1. Let the base BC be bisected by the perp. AD.

Then in $\triangle^s$ ADB, ADC,

Because { BD = DC, [Hyp.]
and AD is common to both;
and ∠ BDA = ∠ CDA, being rt. $\angle^s$;

∴ AB = AC. [I. 4.]

2. Let AD bisect the vert. ∠.

Then in $\triangle^s$ BAD, CAD,

Because { ∠ BAD = ∠ CAD, [Hyp.]
∠ BDA = ∠ CDA, being rt. $\angle^s$;
and AD is common to both;

∴ AB = AC. [I. 26.]

3. Let the base BC be bisected by AD, which also bisects the vert. ∠ BAC.

Produce AD to E, making DE equal to DA, and join CE.

Then in $\triangle^s$ BAD, EDC,

Because { BD = DC, [Hyp.]
AD = DE;
also ∠ ADB = ∠ EDC; [I. 15.]

∴ AB = EC,

and ∠ BAD = ∠ CED. [I. 4.]

∴ ∠ CED = ∠ CAD;

∴ AC = EC.

∴ AB = AC.

4. Let BO, CO, drawn from extremities of the base, meet at O.

Then since OB = OC;
∴ ∠ OBC = ∠ OCB. [I. 5.]
Also ∠ ABO = ∠ ACO; [Hyp.]
∴ ∠ ABC = ∠ ACB;
∴ AC = AB. [I. 6.]

5. Let BD, CE be the equal perps. drawn from ends of base BC.

Then in △s ABD, ACE,

Because { BD = CE, [Hyp.]
∠ ADB = ∠ AEC, being rt. ∠s;
also ∠ at A is common;

∴ AB = AC. [I. 26.]

6. Let CD meet AB in E.

Then in △s ACB, ADB,

Because { AC = AD,
BC = BD, [Hyp.]
and AB is common to both;

∴ ∠ CAB = ∠ DAB. [I. 8.]

Again in △s CAE, DAE.

Because { CA = AD,
AE is common to both,
and ∠ CAE = ∠ DAE;

∴ ∠ AEC = ∠ AED; [I. 4.]
∴ CD is perp. to AB. [Def. 10.]

7. Let BX, CY be drawn perp. to AC, AB respectively, and let them meet in O. Join AO.

Then in △s BXC, CYB,

Because { ∠ BCX = ∠ CBY, [Hyp.]
∠ BXC = ∠ CYB, being rt. ∠s;
and BC is common to both;

∴ △s BXC, CYB are equal in all respects.
and ∠ XBC = ∠ YCB; [I. 26.]

∴ OC = OB. [I. 6.]

Then △ˢ BAO, CAO have their sides respectively equal;

∴ ∠ BAO = ∠ CAO. [I. 8.]

8. In △ˢ BAD, DAE,

Because { ∠ BAD = ∠ DAE, [Hyp.]
∠ BDA = ∠ EDA, being rt. ∠ˢ;
and AD is common;

∴ BD = DE. [I. 26.]

9. In △ˢ ABC, ADC,

Because { AB = AD,
BC = DC, [Hyp.]
and AC is common;

∴ △ˢ ABC, ADC are identically equal; [I. 8.]

∴ ∠ BAC = ∠ DAC, and ∠ BCA = ∠ DCA.

10. In △ˢ ABD, BAC,

Because { BD = AC, [Hyp.]
AB is common,
and AD = BC [Hyp.]

∴ ∠ ABD = ∠ BAC.

∴ △ AKB is isosceles [I. 6]. Similarly △ KDC is isosceles.

11. Here the greatest angle is a right angle. [I. 32.]

Hence the required result follows by Ex. 4 on p. 65.

II. ON INEQUALITIES. Page 101.

1. See Solution of Ex. 5, p. 44.

2. See Solution of Ex. 11, p. 44.

3. See Solution of Ex. 6, p. 55.

4. See Solution of Ex. 9, p. 44.

5. See Solution of Ex. 13, p. 44.

6. See Solution of Ex. 10, p. 44

7. Let O be the given pt., AC and BD the diagonals; then AO + OC > AC, BO + OD > BD [I. 20]. The exceptional case is when O is at the intersection of the diagonals.

8. Let the median AD bisect BC; produce AD to E making DE equal to AD, join EC. Then △s ABD, EDC are identically equal [I. 4] and AB = CE. Now AC + CE > AE [I. 20].

That is, AB + AC > 2AD.

9. This follows at once from Ex. 8, since twice the sum of the sides is greater than twice the sum of the medians.

10. Let the median AD bisect BC. If AD > DC, ∠ ACD is greater than ∠ DAC; similarly ∠ DBA is greater than ∠ DAB. Hence the sum of the angles at B and C is greater than the angle at A; that is, ∠ BAC is acute [I. 32]. The other cases follow similarly.

11. In the rhombus ABCD let ∠ DAB be greater than ∠ ABC. Then since the sides of a rhombus are equal it follows that

DB > AC [I. 25].

13. In the fig. on p. 102 let AD be perp. to BC.

Then ∠ DAC = comp^t. of ∠ ACD,

and ∠ DAB = comp^t. of ∠ ABD;

but ∠ ACD is greater than ∠ ABC;

∴ ∠ DAC is less than ∠ DAB.

∴ ∠ BAD is greater than half vert. ∠ BAC.

∴ AD lies within the ∠ PAC.

Thus by Ex. 12, AP lies between AD and AX, and by Ex. 3 it is intermediate between them in magnitude.

III. On Parallels. Page 103.

2. From O any pt. on the bisector of ∠ BAC draw OP par^l. to AB, and OQ par^l. to AC. Then ∠ QOA = ∠ OAP = ∠ OAQ.

∴ QO = AQ = OP since OPAQ is a par^m. Also OQ = AP; thus the fig. is equilat.

3. Let D be the pt. of intersection of AB and CD; then

$$\angle XYD = \text{alt. } \angle YDA = \angle YDX.$$

$$\therefore YX = DX = XZ \text{ similarly.}$$

4. See Ex. 4, page 60.

5. Let POQ be terminated by the par^ls. at P, Q, and bisected at O; through O draw XOY perp. to the par^ls.; then △^s XOP, YOQ are identically equal [I. 26].

6. The two △^s formed are identically equal [I. 29, I. 26].

7. Let O be pt. equidist. from the par^ls., and let POQ, XOY be drawn to cut them. Draw LOM perp. to the par^ls; then △^s LOP, MOQ are identically equal [I. 26]. ∴ OP = OQ; similarly OX = OY; hence PX = QY [I. 4].

8. Draw XP perp. to CD; bisect ∠ BXP by XQ meeting CD in Q. Through Q draw QY par^l. to XP meeting AB in Y; then Y is the required pt. For

$$\angle YXQ = \angle QXP = \text{alt. } \angle XQY;$$

$$\therefore QY = XY \text{ [I. 6].}$$

9. Bisect ∠ ACB by CD meeting AB in D; draw DE par^l. to BC meeting AC in E. Then ∠ EDC = alt. ∠ DCB = ∠ DCE.

$$\therefore EC = ED \text{ [I. 6].}$$

Again ext. ∠ ADE = int. opp. ∠ ABC = ∠ ACB = ext. ∠ AED [I. 29]; ∴ AD = AE. Hence BD = EC = ED.

10. Bisect the ∠^s ABC, ACB by BO, CO; draw DOE par^l. to BC. Then as in preceding examples it easily follows that

$$DO = DB, \text{ and } EO = EC.$$

11. Produce BC to F. Bisect ∠^s ACF, ABF by CO, BO. Draw OED par^l. to BC meeting AE in E and AB in D. Then as before DO = BD, EO = EC. That is, DE is the diff. between BD and CE.

IV. On Parallelograms. Page 105.

3. In Ex. 2 it is shewn that BC = ZV, and that YZ = YV. Thus BC = 2ZY.

4. In the fig. of Ex. 1 let X, Y, Z be the middle pts. of the sides. Then ZY is par[l]. to BX; similarly XY is par[l]. to BZ; ∴ BZYX is a par[m]., and its diag. ZX bisects it.

5. In fig. of Ex. 2 let ADE be any line meeting ZY in D and BC in E. Then in △ ABE, ZD bisects AE [Ex. 1].

6. In fig. of Ex. 1 let X, Y, Z be middle pts. of sides. Through X, Y, Z draw BC, CA, AB respectively par[l]. to YZ, ZX, XY. Then by I. 34, AZ = XY = BZ.

7. Through P draw PQ par[l]. to AC meeting AB in Q; on QB make QX equal to AQ; join XP and produce it to meet AC in Y. Then QP drawn from middle pt. of AX par[l]. to AY bisects XY [Ex. 1].

8. Let AC meet BX in E and DY in F. Then DY is par[l]. to XB [I. 33], therefore by Ex. 1, CE is bisected by YF, and AF is bisected by XE.

9. Let P, Q, R, S be middle pts. of sides AB, BC, CD, DA respectively. Then by Ex. 2, PQ and SR are each par[l]. to AC, and PS and QR are each par[l]. to BD.

10. In last Ex. PR and QS are diags. of a par[m]. and therefore bisect each other [Ex. 5, p. 70].

11. Let BC, AD be the oblique sides; join BD. Let X, Y be middle pts. of BC, BD; then XY is par[l]. to DC [Ex. 2]. Also XY produced bisects AD [Ex. 1]. Similarly for the other diagonal.

12. As in Ex. 11, let X, Y, Z be middle pts. of BC, BD, AD; then XY = half CD, and YZ = half AB [Ex. 3]. Again, if XYZ meets AC in P, XY = half CD, and XP = half AB; ∴ PY = half diff. of AB and CD.

14. Let three par[l]. st. lines meet a fourth st. line in A, B, C making AB equal to BC, and let them meet another st. line in P, Q, R. Through P draw PST par[l]. to ABC meeting QB in S and RC in T. Then PS = AB = BC = ST [I. 34]; hence PQ = QR [Ex. 1].

15. Let AB, CD be equal and par[l]. st. lines and let XY, PQ be their projections on any st. line; let AE, CF drawn par[l]. to XY meet BY, DQ in E and F respectively. Then △[s] ABE, CDF are identically equal [I. 26], so that AE = CF. ∴ XY = PQ [I. 34].

16. Let OZ be perp. to XY. Then XZ = ZY being projections of the equal lines AO, OB. ∴ the △s XZO, YZO are identically equal [I. 4].

17. Draw ALM par^{l}. to XY meeting OZ in L, BY in M. Then BM = 2OL [Ex. 1, 3, p. 104]. Also AX = LZ = MY.

∴ 2OZ = 2OL + 2LZ = BM + MY + AX = BY + AX.

18. The first case can be proved as in Ex. 17. In the second case, with same construction as before, BN = OL = NM.

∴ 2OZ = 2OL − 2LZ = BM − MY − AX = BY − AX.

20. Let ABCD be the given par^{m}. Through A draw any st. line EAF and let CX, BF and DE be perps. on this line. Through C draw CH par^{l}. to EF meeting FB in H. Then it is easily seen that △s BCH, DAE are identically equal [I. 26]; ∴ BH = DE.

∴ DE + BF = BH + BF = CX [I. 34],

for CXFH is a par^{m}. by constr.

21. Let AX, CY be perps. on the given line from one pair of opp. ∠s, and DP, BQ perps. from the other pair of opp. ∠s. Let the diagonals intersect in E, and let EF be perp. to the given line.

Then AX + CY = 2EF [Ex. 17, p. 106]

= DB + BQ,

since E is the middle pt. of the diagonals [Ex. 5, p. 70].

22. From D in base BC let DE, DF be drawn perp. to AC, AB respectively; from B let BG be drawn perp. to AC. Draw BH par^{l}. to AC to meet ED produced in H. Then GH is a par^{m}. and BG = EH. Also △s BFD, BHD are identically equal [I. 26], so that DH = DF. That is, BG = sum of DE and DF.

23. Take D in CB produced, then with the same lettering and construction as in Ex. 22 it is easily seen that BG = HE = difference between DE and DF.

24. Let OX, OY, OZ be perps. to BC, CA, AB respectively. Through O draw POQ par^{l}. to BC; then APQ is an equilat. △ and sum of OY and OZ = perp. from P on the opp. side = perp. from A on PQ since △ APQ is equilat. Hence sum of OX, OY, OZ = perp. from A on BC.

25. Let O be the given pt., AB, CD the par[l]. lines. With A as centre and radius equal to given length describe a circle. This will in general cut CD in two pts. L, M. Then lines drawn through O par[l]. to AL and AM will be the required lines.

26. Let AB be the line to which the required line is to be par[l]., QP and RS the other two given lines. Let QP meet AB in P; draw PT on AB equal to the given length; through T draw TR par[l]. to PQ meeting SR in R; through R draw RQ par[l]. to AB meeting PQ in Q. Then QR is the required line [I. 33].

27. Let the given lines PO, QO meet in O; bisect ∠ POQ by OS. Draw OR perp. to OS on the same side as OQ and equal to the given length; through R draw RT par[l]. to OP meeting OQ in T. Through T draw TV par[l]. to OR meeting OP in V. Then TV is equal and par[l]. to OR [I. 33], and since it is perp. to OS it is equally inclined to OP and OQ [I. 26].

28. Join AP; bisect AP in Q, and draw QR par[l]. to AB (the further line) meeting AC in R. Join PR and produce it to meet AB in S. Then QR bisects PS [Ex. 1, p. 104].

V. Miscellaneous Theorems and Examples. Page 108.

1. ∠ ACD = ∠ ADC [I. 5] and ∠ ACB = ∠ ABC; ∴ ∠ BCD = sum of ∠[s] CBD, BDC. That is, ∠ BCD is a rt. ∠ [I. 32].

2. Let CD join the rt. ∠ C to D the middle pt. of AB. Draw DE par[l]. to AC meeting BC in E. Then BC is bisected at E [Ex. 1, p. 104]. Also ∠[s] DEB, DEC are equal, being rt. ∠[s].

∴ △[s] DEB, DEC are identically equal [I. 4].

∴ DC = DB.

3. By Ex. 2 each of the lines is equal to half the base.

4. By Ex. 2, p. 96, AZYX is a par[m]. ∴ ZXY = ∠ BAC.

Again DY = AY in the rt. angled △ ADC [Ex. 2];

∴ ∠ YDA = ∠ YAD.

Similarly ∠ ZDA = ∠ ZAD; ∴ ∠ ZDY = ∠ BAC.

5. Let AD be the perp. on the hypotenuse. Then

∠ DAC = comp[t]. of ∠ DCA = ∠ ABC.

8. This follows at once from Ex. 7 (ii) and Ex. 3, p. 65.

9. The ∠ at B = diff. of ∠s BCD, BAC [I. 32],

and ∠ at F = diff. of ∠s FCD, FAC

= half diff. of ∠s BCD, BAC (hyp.).

10. Let ∠ B be double of ∠ A. Let CD be drawn from rt. ∠ to middle pt. of hypotenuse AB. Then since CD = DA, ∠ CDB is double of ∠ CAD. ∴ ∠ CBD = ∠ CDB, so that CB = CD = half hypotenuse [Ex. 2, p. 108].

11. Let ABCD be a par^{m}. and let BE = DF on the diag. BD. Then △s ABE, CDF are identically equal [I. 4], so that AE = FC. Also ∠ CFE = supp^{t}. of ∠ DFC = supp^{t}. of ∠ AEB = ∠ AEF. ∴ FC is par^{l}. to AE; ∴ AF is equal and par^{l}. to EC [I. 33].

12. The △s ACZ, ABX are identically equal [I. 4].

∴ ∠ ZAR = ∠ ACZ.

Now ext. ∠ PRQ = sum of ∠s RAC, ACR = sum of ∠s RAC, ZAR = ∠ BAC. Similarly each of ∠s of △ PQR may be proved equal to the angle of an equilat. △.

13. The △s APS, CRQ are identically equal [I. 4].

∴ PS = QR, and ∠ APS = ∠ QRC.

Again ∠ APR = alt. ∠ CRP;

∴ ∠ SPR = ∠ PRQ.

That is, SP is equal and par^{l}. to QR. Hence SR is equal and par^{l}. to PQ [I. 33].

14. It may be proved as in the prop. that △ ABF is equilat. ∴ ∠ CAF = two-thirds of 2 rt. ∠s, and ∠ PAF is one-third of 2 rt. ∠s. Again AP = AF. ∴ ∠ APF = AFP, and each is one-third of 2 rt. ∠s [I. 32]. Similarly ∠ BFQ is one-third of 2 rt. ∠s. Thus the three ∠s at F together = 2 rt. ∠s. Also ∠s at P and Q being each equal to ∠ C, △ CPQ is equilat.

15. Let AB be the given st. line, P and Q the given points. At A and B make ∠s BAC, ABD each equal to the angle of an equilat. △. Through P and Q draw st. lines par^{l}. to AC, BD meeting AB in X and Y and intersecting in Z. Then XYZ is the required △.

16. Let O be the given pt., AB and CD the two given st. lines of which AB is the nearer to O. Draw OEF perp. to AB, CD respectively, and OG perp. to OF making OG equal to OF. Draw GH perp. to AB; join OH, and draw OK perp. to OH meeting BC in K. Then △^{s} OHG, OFK are identically equal [I. 26], and OH = OK.

The line OG may be drawn par^l. to AB in either direction; thus there will be two solutions corresponding to each position of O.

17. Let AB, AC, AD be the three given st. lines. Take any pt. P in AD; draw PQ par^l. to AB meeting AC in Q, and draw QR par^l. to AD. Then APQR is a par^m. and its diagonals bisect each other [Ex. 5, p. 70]. Thus PR is bisected by AQ. As P may be taken anywhere on AD the number of solutions is unlimited.

18. Let L, M, N be the three given lengths, and B the given point. From B draw BC equal to N; and on BC describe a △BFC, having BF equal to twice M and CF equal to L. Bisect BF at E. Join CE, and produce it to A, making EA equal to CE. Join BA. Then BA, BE, BC are the required lines. For BC = N, and BE = M by constr., and it may be shewn that the △^{s} AEB, CEF are identically equal [I. 4]. ∴ AB = CF = L. Also AE = EC by constr.

19. Let ABC be an equilat. △; bisect the angles at B and C by BO, CO; through O draw OD, OE par^l. to AB and AC respectively meeting BC in D and E. Then, by I. 29, 32, △ODE is equiangular to △ABC, so that ODE is equilat. Again

∠ DOB = alt. ∠ ABO = ∠ OBD;

∴ OD = BD.

Similarly OE = OC.

∴ BD = DE = EC.

20. Bisect ∠ ABC by BO meeting AC in O; through O draw OD par^l. to AB and OE par^l. to BC meeting BC and AB in D and E respectively. Then as in Ex. 19, OD = BD, and OE = BE. Hence it easily follows that EBDO is a rhombus [I. 34].

VII. On the Construction of Triangles with given parts.

Page 115.

2. Let the given diff. be equal to AB and the hypotenuse equal to K. From A draw AE making with BA produced an ∠ equal to half a rt. ∠. From centre B, with radius equal to K, describe a circle cutting AE or AE produced in the points C, C′. From C and C′ draw perps. CD, C′D′ to AB; and join CB, C′B. Then either of the △ˢ CDB, C′D′B will satisfy the given conditions. [See Note to Ex. 1.]

3. See fig. on p. 97. Let AB be the given sum, then using the construction and proof given on p. 97, it is shewn that AX = XC, and BX = BC. Thus CBX is the required triangle.

5. This is clearly a particular case of the preceding example.

6. Let P, Q be the given pts. through which the sides are to pass, XY the st. line in which the base is to be. At X draw XZ equal to the given altitude. Through Z draw ZKL parˡ. to XY. Then by Ex. 7, page 55, draw through PQ two lines PK, QK making equal angles with ZKL. Produce KP, KQ to meet XY in M and N; then KMN is the required △.

7. Draw AB, CD two parˡ. st. lines at a dist. from each other equal to the given altitude. At P, any pt. in AB, make ∠ APQ = one-third of two rt. ∠ˢ [I. 1] and ∠ BPR = one-third of two rt. ∠ˢ. If PQ, PR meet CD in Q and R respectively, PQR is the required △.

8. Let AB be the base and K the given difference. Bisect AB at X; from X draw XH perp. to AB, making XH equal to K; join AH. At the pt. A make ∠ HAC equal to ∠ AHX, and since HX < AX [Ex. 7, p. 44], ∴ C must fall on the side of AB which is remote from H. Let AC meet HX produced in C; join CB. Then ACB is the required △. [See proof of Ex. 1, p. 96.]

9. Let AB be the base; make ∠ BAX equal to given ∠, and AX equal to sum of sides. Join BX. At B in XB make ∠ XBC equal to ∠ BXA. Then ACB is the required △.

10. Let AB be the base; make ∠ BAX equal to given ∠, and AX equal to diff. of sides. Join BX. Produce AX to C, and make ∠ XBC equal to ∠ BXC. Then ACB is the required △.

12. Let AB be the given base, K the sum of the remaining sides and X the difference of the ∠ˢ at the base. Make the ∠ ABD equal to half the ∠ X; draw BE perp. to BD, and from centre A and with radius equal to K describe a circle cutting BE in E. At B make ∠ EBC equal to ∠ AEB. Then ACB is the required △. [Ex. 7, p. 109.] Since, if AE meets BD at F, it may be shewn that CB = CF.

13. Let AD be the given perp. and let the two given differences be X and Y. On AD as base describe a △ ABD having ∠ ADB a rt. ∠ and the diff. of AB and BD equal to X. Also on the other side of AD describe a △ ADC having ∠ ADC a rt. ∠ and the diff. of AC and DC equal to Y. [Ex. 10, p. 116.]

Then ABC is the required △.

VIII. On Areas. Page 117.

1. Let ABCD be a parᵐ., O the middle pt. of the diag. BD. Draw any line through O meeting AB, CD in E and F respectively. Then △ˢ EOB, DOF are identically equal [I. 29, 26].

∴ AEFD = △ ADB = half the parᵐ.

2. Join the given pt. to the middle pt. of a diagonal, and produce it to meet two of the parallel sides.

Examples 3 and 4 are particular cases of Ex. 1.

5. Let EXF drawn parˡ. to AD meet DC in E and AB in F. Then △ˢ BXF, XEC are identically equal [I. 29, 26]. ∴ the area of the trapezium is equal to that of parᵐ. ADEF.

6. In the preceding Example DE = half the sum of DE and AF, that is half the sum of DC and AB, since BF = EC.

7. For △ AXD is half the parᵐ. ADEF in Ex. 5.

8. Let E, F be middle pts. of AB and DC; join ED, EC. Then △ˢ EDF, EFC are equal, and △ˢ AED, BEC are equal [I. 38].

9. The △ˢ ADB, BCD are equal [I. 37]. Take away the common part, the △ BXD.

10. The $\triangle^s$ ADB, BCD are equal, and therefore AC is par[l]. to BD [I. 39].

11. This may be proved by considering $\triangle^s$ BCX, DCX on a common base CX and of equal altitudes [Ex. 13, p. 70]. Or if the diagonals meet in O, the $\triangle^s$ OBX, ODX are equal, and $\triangle^s$ ODC, OBC are equal [I. 38].

12. $\triangle$RBC = $\triangle$QBC [Ex. 2, p. 104 and I. 37];

$\therefore$ $\triangle$RBX = $\triangle$QCX.

Again $\triangle$ABQ = $\triangle$BQC; $\therefore$ by taking away the equal $\triangle^s$ RBX, QCX, the area AQXR = $\triangle$BXC.

13. Let ABCD be the quad[l]., and P, Q, R, S the middle points of the sides AB, BC, CD, DA.

Draw AC, BD intersecting at O. Let AO meet PS at X. Then PS is par[l]. to BD, and AX = XO. [Ex. 1 and 2, p. 104.]

First shew that the perps. from A and O on PS are equal [I. 26]. Hence it follows that the $\triangle$APS = the $\triangle$POS.

Similarly $\triangle$BPQ = $\triangle$POQ, $\triangle$QCR = $\triangle$QOR, and $\triangle$SDR = $\triangle$SOR.

Hence by addition the par[m]. PQRS is one half of the quad[l].

14. Let C and D be vertices of two equal $\triangle^s$ ACB, ADB on opposite sides of the common base AB; let CD meet AB or AB produced in E. Then if DF, CG are drawn perp. to AB, DF = CG, and the $\triangle^s$ DEF, CEG are identically equal [I. 26].

15. Let ABCD be the trapezium having AB par[l]. to DC. Bisect BD, CA in E and F and join EF. Then $\triangle^s$ AEB, AFB are equal, being halves of equal $\triangle^s$ ADB, ACB [I. 38].

$\therefore$ EF is par[l]. to AB [I. 39].

16. (i) Let ABC, DBC be two $\triangle^s$ on a common base BC, ABC having the greater altitude.

Draw BX perp. to BC, and through A and D draw AA′, DD′ par[l]. to BC to meet BX at A′, D′. From A′X cut off A′E equal to BD′, and join EC, A′C, D′C.

Then EB is the sum of the altitudes of the $\triangle^s$ ABC, DBC.

And since EA′ = D′B, $\therefore$ $\triangle$EA′C = $\triangle$D′BC [I. 38].

Hence △ ABC + △ DBC = △ A′BC + △ D′BC

= △ A′BC + △ EA′C

= △ EBC.

(ii) may be proved by a similar method.

17. If O lies between the parallel sides AB, CD, the perp. EOF to these sides is equal to the perp. from A to CD.

Thus the △s OAB, OCD, ADC have equal bases, and the altitude of ADC is equal to the sum of the altitudes of the other two.

∴ the sum of the △s OAB, OCD = △ ADC which is half the par^m^.

If O does not lie between AB and CD, the diff. of the △s OAB, OCD = △ ADC.

18. (ii) If O is within ∠ DAB and its opp. vert. ∠, then AO intersects the par^m^.; so that the perp. from C on OA is equal to the diff. of the perps. drawn from B and D to OA [Ex. 20, p. 107]. Therefore since the △s OAC, OAD, OAB are on the same base OA, △ OAC = diff. of △s OAD, OAB.

19. Let the lines EOF, GOK drawn through O par^l^. to AD, AB respectively meet AB in F, AD in G.

Then par^m^. GB = 2 △ AOB, and par^m^. DF = 2 △ DOA. And since par^m^. GF is common to these two par^ms^., the diff. between par^ms^. BO and DO = twice the diff. between △s AOB and DOA = 2 △ AOC [Ex. 18. ii.].

20. Draw BO par^l^. to the diag. AC, and CO par^l^. to AB; then ABOC is a par^m^. Also the perp. from D on BO is equal to the sum of perp. from D on AC and perp. from B on AC.

∴ △ DBO = △ DAC + △ ABC,

since these △s have equal bases [Ex. 16 (1)].

22. Let ABC be the given △, and B the given ∠. In BA take BD equal to the base of the required △, and by Ex. 21 draw through D a st. line to meet BC produced in E, so that △ DBE may be equal to △ ABC.

23. Let CAB be the given △ on base AB. Through A draw AD perp. to AB and equal to the given altitude, and through C draw CE par^l^. to AB meeting AD in E. Join DB, and draw EF par^l^. to DB meeting AB in F. Then △ DAF = △ EAB = △ CAB [Ex. 21].

24. Let AB be the given base, CD the given line in which the vertex is to lie. On AB describe a △ABE equal to the given △ [Ex. 21]. Through E draw EC par^{l}. to AB meeting CD in C; then CAB is the required △.

25. (i) On AB the given base describe △ABC equal to the given △ [Ex. 21, p. 119]. Bisect AB at D, draw DE perp. to AB meeting CE, drawn par^{l}. to AB in E; then AEB is the req^{d}. △.

(ii) Draw AF perp. to AB meeting CF, drawn par^{l}. to AB in F; then FAB is the req^{d}. △.

26. Let X, Y be the two given △^{s}. If they are not on equal bases, make a triangle Z equal to Y and having a base equal to that of X [Ex. 21, p. 119]. The construction now follows easily by Ex. 16 (i), p. 118.

27. Through A draw AD par^{l}. to BC meeting BX in D. Then △CDB = △ABC. Through X draw XF to meet BC so that

△XBF = △CDB [Ex. 21, p. 119].

28. The △^{s} BDX, BDC are equal [I. 37]. To each add △ABD.

29. Take a five-sided figure ABCDE. Join EC. Through D draw DF par^{l}. to EC meeting BC at F. Then the quadrilat. ABFE = the given fig. For △EFC = △EDC [I. 37]; add to each the fig. ABCE. Similarly a six-sided figure can be replaced by an equal figure of five sides, and so on. Thus any rectilineal figure can ultimately be reduced to a triangle of equal area.

30. Through C and D draw CE, DF par^{l}. to BX and AX respectively meeting AB in E and F; then EXF is the req^{d}. △ [Ex. 21, p. 119].

31. Let ABCD be the par^{m}.; through C draw CE par^{l}. to the diag. BD; bisect BD at F, and draw FG perp. to BD meeting CE in G; join GB and GD. Then it is easily seen that BGD is an isosceles △ equal in area to △BCD, and by drawing a △BHD identically equal to △BGD but on the opp. side of BD, a rhombus BGDH will be formed equal to the given par^{m}.

32. Let ABC be the given △ on BC as base. Bisect BC in D, and draw AE par^{l}. to BC. With centre B and radius equal to half the sum of sides BA, AC, describe a circle cutting AE in E;

through D draw DF par^{l}. to BE meeting AE in F. Then par^{m}. BEFD is double of △ABD, and is therefore equal to △ABC. Also sum of BE and DF = sum of BA and AC, and sum of BD and EF = BC.

33. The bisecting line is the median through the given angular pt. [I. 38].

34. Join the given pt. to the pts. of trisection of the opp. side.

35. Divide the side opp. to the given pt. into the required number of parts, and join the points of division to the given pt.

38. As the method is quite general it will be sufficient to take a particular case. Let ABC be a triangle from which it is required to cut off a fifth part by a st. line through a pt. X in AB. Take BD a fifth part of BC [Ex. 19, p. 107]. Join AD, and through X draw XE to meet BC in E, so that △BXE = △ABD [Ex. 21, p. 119].

40. With the construction of Ex. 28, p. 120 let BAX be a △ equal to the given quadrilateral. Take AY equal to one-fifth of AX; join BY. Then △BAY = one-fifth of △BAX, that is, of the quadrilateral. The method is quite general.

41. (i) Let AL meet BC in X. Then

sq. on AB = sum of sqq. on BX, AX
= sum of sqq. on DL, AX;
sq. on AE = sum of sqq. on AL, LE
= sum of sqq. on AL, CX.

∴ sum of sqq. on AB, AE = sum of sqq. on DL, AL, AX, CX = sum of sqq. on AD, AC.

(ii) Produce AC to M making CM equal to AC; join BM. Then ∠BCM = supp^{t}. of ∠ACB = ∠ECK, and △^{s} BCM, ECK are identically equal [I. 4]. Therefore

sq. on EK = sq. on BM
= sum of sqq. on BA, AM [I. 47]
= sq. on BA with four times sq. on AC.

(iii) Sq. on EK = sq. on AB, with four times sq. on AC,

sq. on FD = sq. on AC, with four times sq. on AB;

∴ sum of sqq. on

EK, FD = five times sq. on AB with five times sq. on AC

= five times sq. on BC.

42. Let AB be divided at C; from C draw CD perp. to AB and equal to CB. Join AD, DB. Then ∠ CDB = ∠ CBD [I. 5]; ∴ ∠ ADB is greater than ∠ ABD, and ∴ AD < AB.

Now sq. on AD = sum of sqq. on AC, CD [I. 47]

= sum of sqq. on AC, CB.

∴ sq. on AB is greater than sum of sqq. on AC, CB.

43. Let sq. on AC be greater than the sum of sqq. on AB, BC; then shall ∠ ABC be obtuse. Draw BD perp. to BC and equal to AB.

Then sq. on DC = sum of sqq. on DB, BC [I. 47].

But sq. on AC > sum of sqq. on AB, BC.

∴ sq. on AC > sq. on DC, or AC > DC.

Hence by I. 24 in the △s ABC, DBC, the ∠ ABC is greater than ∠ DBC; that is, ∠ ABC is obtuse.

44. The sq. on BQ = sum of sqq. on AB, AQ [I. 47];

and sq. on CP = sum of sqq. on AC, AP.

∴ sum of sqq. on BQ, CP = sum of sqq. on AP, AQ, together with sum of sqq. on AB, AC.

That is, sum of sqq. on BQ, CP = sum of sqq. on PQ, BC.

45. Let the medians be BQ, CP. Then by preceding example four times the sum of the sqq. on BQ, CP = four times the sum of the sqq. on PQ, BC.

But four times the sq. on PQ = sq. on BC [Ex. 3, p. 105]. ∴ four times the sum of the sqq. on BQ, CP = five times the sq. on BC.

46. Let AB be a side of the given square; produce AB to C making BC equal to AB. On AC describe an equil. △ ACD. Join DB. Then, as in I. 11, DB is perp. to AC, and therefore the sq. on AD = sq. on AB with sq. on DB. But AD is double of AB:

∴ sq. on AD = 4 times sq. on AB.

∴ sq. on DB = 3 times sq. on AB.

47. Let AB be the st. line to be divided, K a side of the given sq. At B make ∠ ABD equal to half a rt. ∠. From centre A with radius K draw a circle to cut BD at C and C′. From C (or C′) draw CX perp. to AB. Then AB is divided as required at X.

For sq. on AC = sqq. on AX, XC [I. 47]
= sqq. on AX, XB,

for XC = XB, since CXB is a rt. ∠, and XBC is half a rt. ∠ [I. 32, and I. 6].

There will be two solutions, one solution, or no solution, according as the circle with radius K cuts BC in two pts., touches at one, or does not meet it at all.

IX. On Loci. Page 125.

4. The locus is a concentric circle whose radius is equal to the sum or difference of the radius of given circle and the given distance.

5. Let OA, OB be the two intersecting st. lines, K the given const. length. At O draw OC perp. to OA and equal to K; draw CD par[l]. to OA meeting OB in D. In OA make OE equal to OD and join DE. Then DE is the required locus; for by Ex. 22, p. 107, the sum of the perps. on OA and OB from any pt. in DE is equal to the perp. from D on OA = OC = K.

6. With the same lettering and construction as in Ex. 5, let DE be produced both ways indefinitely to X and Y. Then the required locus is the part of XY external to the △ODE. [See Ex. 23, p. 107.]

7. Let AB be the rod of given length, and C the intersection of the rulers at right angles. Bisect AB in D; then CD = half of AB. Therefore the locus of D is a circle whose centre is the fixed pt. C and whose radius is half the given length of the rod.

8. Let ACB be a rt. angled △ on AB as hypotenuse. Then, if AB is bisected at D, CD is constant, being equal to half of AB. Thus the locus is a circle whose centre is D and radius equal to half the given base.

9. Bisect AB in C and AX in D and join DC. Then the locus required is the locus of the vertices of right angled △s on base AC as hypotenuse [Ex. 8].

10. Let AB be the base, and AD the altitude, which is known since the base and area are given. Then the locus required is a st. line through D par[l]. to AB [I. 39].

11. The diagonals of a par[m]. divide it into four equal △[s] having their vertices at the inters[n]. of the diagonals. Thus the problem is the same as in Ex. 10, and the required locus is a st. line through the inters[n]. of the diagonals par[l]. to the given base.

12. Let BC be the given base, and ABC the △ in any of its positions. Then since the area of the △ is constant, A must move on a fixed st. line par[l]. to BC. Let AX be the median bisecting BC; then if O is the inters[n]. of medians AO = 2OX [Ex. 4, p. 113]. Draw AD perp. to BC and in it take AP = 2PD [Ex. 19, p. 107]. Join OP. Then it may be shewn [as in Ex. 2, p. 104] that OP is par[l]. to BC. But P is clearly at a *fixed distance from* BC (being one-third of the distance between the par[ls].); ∴ O lies on a st. line par[l]. to BC and at a fixed distance from it.

X. Intersection of Loci. Page 127.

1. Let X, Y be the given pts., AB the given st. line. Join XY and bisect it in C; draw CZ perp. to XY meeting AB in D. Then from equal △[s] DCX, DCY [I. 4] DX = DY.

2. If the lines are not par[l]. let them be AB, CB meeting in B; and let P, Q be the given distances. Draw BE perp. to AB and equal to P; at E draw EF par[l]. to BA; then the required pt. lies in EF.

Again draw BD perp. to BC and equal to Q; at D draw DF par[l]. to BC; then the required pt. lies in DF. That is, F is the required pt.

There are four solutions since each of the lines BE and BD may be drawn in either direction.

3. Let AB be the given base, X the given ∠, and Y the length of the side opposite. At A draw AZ making ∠ BAZ equal to X. From B as centre and with radius equal to Y describe a circle. Draw BD perp. to AZ; then BD is the shortest distance of B from AZ.

(i) If $Y < BD$, the circle will not cut AZ, and there is no △ possible with the given parts.

(ii) If $Y = BD$, the circle will meet AZ in one pt., and there is one solution, viz. the right-angled △ BAD.

(iii) If $Y > BD$, the circle will cut AZ in two pts. C_1, C_2 and there will be two triangles ABC_1, ABC_2 each of which satisfy the data. This last case however requires that Y shall be less than AB, for then both pts. C_1, C_2 will lie on AZ on the same side of A. If one of these pts. C_2 lies on opposite side, the △ BAC_2 would have the ∠ BAC_2 equal to the supplement of given ∠ X, and would not satisfy the data.

4. Let ABC be the given △ on base BC, and DE the given st. line. Through A draw AF par[l]. to BC meeting DE in F and join FB, BC. Then FBC is the required △ [I. 37]. If ED is par[l]. to BC the solution is only possible when DE passes through A. In this case any pt. in DE may be taken as the vertex of the required △, and the number of solutions is unlimited.

5. Let ABC be the given △ on base BC. Draw AD par[l]. to BC. Bisect BC at E and draw ED perp. to BC meeting AD in D. Then BDC is the required △ [I. 4 and I. 37].

6. Let AB, CD be the two given par[l]. st. lines, and X the given pt. Draw AC perp. to CD and bisect it at E. Through E draw EF par[l]. to either of the given st. lines. Then the required pt. must lie in this st. line. With centre X and radius equal to the given distance describe a circle cutting EF in P and Q; then P and Q satisfy the required conditions. If the circle meets EF in one pt. only there is one solution; if the circle does not meet EF there is no solution.

BOOK II.

EXERCISES.

Page 137.

(i) $AP = \frac{1}{2}AB = \frac{1}{2}\{AQ + QB\}$.

(ii) Mark off AQ′ equal to BQ. Then P is clearly the middle point of QQ′, so that

$$PQ = \frac{1}{2}QQ' = \frac{1}{2}(AQ - AQ') = \frac{1}{2}(AQ - BQ)$$

Page 139.

Take the enunciations as given on p. 139, that is,

$$AQ \,.\, QB = PB^2 \sim PQ^2$$

$$= \left\{\frac{AQ + QB}{2}\right\}^2 \sim \left\{\frac{AQ - QB}{2}\right\}^2.$$

(See the last Exercise.)

Page 151.

(i) The two $\triangle^s$ FAB, HAC are identically equal [I. 4].

$\therefore$ the $\angle$ ABF = the $\angle$ ACH,

and the $\angle$ LHB = the vert. opp. $\angle$ AHC;

$\therefore$ the $\angle^s$ HBL, LHB = the $\angle^s$ ACH, AHC,

$\therefore$ the remaining $\angle$ HLB = the remaining $\angle$ HAC [I. 32]

= a rt. angle.

(ii) Since EF = EB, ∴ ∠ EBF = ∠ EFB.

And in the △ˢ OBL, CFL, we have the ∠ˢ OLB, CLF rt. angles,

∴ the ∠ FCL = the ∠ BOL = the ∠ EOC [I. 15].

∴ EO = EC = EA.

Hence it may be proved that AOC is a rt. angle after the method of III. 31.

(iii) Produce FG, DB to meet at M. Then since the sq. FH = the rect. HD, these are complementary parᵐˢ., and hence H lies on the diagonal CM. [I. 43. *Converse.*]

Hence it may easily be shewn that the diagonals GB, FD, AK of the *rectangles* GHBM, FCDM, ACKH are parˡ.

(iv) For the fig. FK = the fig. AD;

or, the rect. CF, FA = the sq. on AC.

Page 155.

1. Let ABC be an isosceles △, having its vertex at A, and let BD be drawn perp. to AC.

Then each of the ∠ˢ ABC, ACB must be acute. [I. 17.]

Hence by II. 13, $AB^2 = AC^2 + BC^2 - 2AC \cdot CD$.

But $AB^2 = AC^2$, so that $BC^2 = 2AC \cdot CD$.

2. Let ABC be the △, having the ∠ ABC equal to one-third of two rt. angles. From A draw AX perp. to BC. First prove AB = 2 . BX: this may be done by joining X to the middle point of AB. [See Ex. 4, p. 65.]

Then by II. 13, $AC^2 = AB^2 + BC^2 - 2BX \cdot BC$

$= AB^2 + BC^2 - AB \cdot BC$.

3. Let the △ ABC have the ∠ ABC equal to two-thirds of two rt. angles. Draw AX perp. to CB produced.

Then the ∠ ABX = one-third of two rt. angles; hence, as in the last Ex., AB = 2BX.

And by II. 12, $AC^2 = AB^2 + BC^2 + 2BX \cdot BC$

$= AB^2 + BC^2 + AB \cdot BC$.

THEOREMS AND EXAMPLES ON BOOK II.

On II. 4 and 7.

1. Let AB be the st. line, and C its middle point, then by II. 4

$$AB^2 = AC^2 + CB^2 + 2AC \,.\, CB$$
$$= AC^2 + AC^2 + 2AC \,.\, AC$$
$$= AC^2 + AC^2 + 2AC^2 = 4AC^2.$$

2. Let AB be divided into three parts at the points X, Y.

On AB describe the sq. ABCD. Join BD. Through X and Y draw XP, YQ par[l]. to AD or BC and cutting BD at H and K. Through H and K draw LHMN and EFKG par[l]. to AB. Then prove as in II. 4 that

(i) the figs. LP, FM, YG are respectively the sqq. on AX, XY, YB.

(ii) the fig. AF = the fig. MC = the rect. AX, YB.

(iii) the fig. EH = the fig. HQ = the rect. AX, XY.

(iv) the fig. XK = the fig. KN = the rect. XY, YB.

3. Let ABC be a △ rt. angled at B, and let BD be drawn perp. to AC.

Then by II. 4, $AC^2 = AD^2 + DC^2 + 2AD \,.\, DC.$

But by I. 47, $AC^2 = AB^2 + BC^2$

$$= AD^2 + BD^2 + DC^2 + BD^2.$$

$$\therefore\ AD^2 + DC^2 + 2BD^2 = AD^2 + DC^2 + 2AD \,.\, DC,$$

or $BD^2 = AD \,.\, DC.$

4. Let ABC be an isosceles $\triangle$, having its vertex at A. Draw BD perp. to AC.

Now by II. 4, $AC^2 = AD^2 + DC^2 + 2AD.DC$.

Add to each of these equals BD^2,

$$\therefore\ AC^2 + BD^2 = AD^2 + BD^2 + DC^2 + 2AD.DC,$$

or, $$AC^2 + BD^2 = AB^2 + DC^2 + 2AD.DC.$$

But $$AB^2 = AC^2;$$

$$\therefore\ BD^2 = DC^2 + 2AD.DC.$$

5. Let ABCD be a rectangle; and on AB, BC let sqq. ABGH, BCEF be described externally to the rectangle.

Join HB, BE. Then HB, BE are clearly in one line, for the $\angle$s ABH, CBE are each half of a rt. angle.

Also as in II. 9, $HE^2 = 2HD^2$.

But by II. 4, $HE^2 = HB^2 + BE^2 + 2HB.BE$,

and $$HD^2 = HA^2 + AD^2 + 2HA.AD.$$

So that $$HB^2 + BE^2 + 2HB.BE = 2HA^2 + 2AD^2 + 4HA.AD.$$

But, as in II. 9, $HB^2 = 2HA^2$, and $BE^2 = 2BC^2 = 2AD^2$,

$$\therefore\ 2HB.BE = 4HA.AD;$$

or, the rect. HB . BE = twice the rectangle ABCD.

6. Let ABC be the given $\triangle$. From A draw AX perp. to BC.

Then by II. 4, $BC^2 = BX^2 + XC^2 + 2BX.XC$.

To each of these equals add $2.AX^2$.

So that $$BC^2 + 2AX^2 = BX^2 + AX^2 + XC^2 + AX^2 + 2BX.XC$$
$$= AB^2 + AC^2 + 2BX.XC. \quad [\text{I. } 47.]$$

On II. 5 and 6.

7. Let ABC be a $\triangle$ rt. angled at B.

Then $$AB^2 = AC^2 - BC^2. \quad [\text{I. } 47].$$
$$= (AC + BC)(AC - BC). \quad [\text{II. } 5.]$$

9. Let ABC be an isosceles △, having its vertex at A; and let AQ be drawn to any point Q in BC.

Draw AP perp. to BC. Then, by I. 26, BP = PC.

Hence by II. 5, rect. BQ, $QC + PQ^2 = PC^2$.

To each of these equals, add AP^2; so that

$$BQ . QC + PQ^2 + AP^2 = PC^2 + AP^2;$$

or, $$BQ . QC + AQ^2 = AC^2. \quad [\text{I. } 47.]$$

10. Let ABC be the isosceles triangle having its vertex at A. and let Q be any point in the base BC produced.

Join AQ, and draw AP perp. to BC.

Then it may be shewn by I. 26 that BC is bisected at P.

Hence by II. 6, $$BQ . QC + PC^2 = PQ^2.$$

To each of these equals add AP^2.

$$\therefore BQ . QC + PC^2 + AP^2 = PQ^2 + AP^2;$$

or, $$BQ . QC + AC^2 = AQ^2.$$

11. Taking the same letters as in the last Ex. we may shew that $$BQ . QC + AC^2 = AQ^2.$$

But $$BQ . QC = AC^2 \quad [\text{Hyp.}]$$

$$\therefore AQ^2 = 2AC^2.$$

12. Draw XP, YQ perp. to BC.

Then $$BY^2 - YC^2 = BQ^2 - QC^2 \quad [\text{Ex. 8, p. 159}]$$
$$= (BQ + QC)(BQ - QC) \quad [\text{II. } 5]$$
$$= BC . PQ = BC . XY.$$

13. Let ABC be the △ rt. angled at B.

Draw BX perp. to AC.

Then by II. 7, $$AC^2 + CX^2 = 2AC . CX + AX^2.$$

But $$AC^2 = AB^2 + BC^2 = 2BX^2 + AX^2 + XC^2 \quad [\text{I. } 47].$$

$$\therefore 2BX^2 + 2CX^2 + AX^2 = 2AC . CX + AX^2;$$

or, $$BX^2 + CX^2 = AC . CX;$$

or, $$BC^2 = AC . CX. \quad [\text{I. } 47.]$$

On II. 9 and 10.

14. Let AB be the given st. line divided equally at P and unequally at Q.

Then $AB^2 = 4AP^2$; also $AB^2 = AQ^2 + QB^2 + 2AQ . QB$. [II. 4.]

Hence $AQ^2 + QB^2 = 4AP^2 - 2AQ . QB$.

But $AQ . QB = AP^2 - PQ^2$ [II. 5].

$\therefore AQ^2 + QB^2 = 4AP^2 - 2AP^2 + 2PQ^2$;

or, $AQ^2 + QB^2 = 2\{AP^2 + PQ^2\}$.

15. Let the st. line AB be bisected at P and produced to Q.

Then, as before, $AB^2 = 4AP^2$;

also by II. 7, $AQ^2 + QB^2 = 2AQ . QB + AB^2$.

$\therefore AQ^2 + QB^2 = 2AQ . QB + 4AP^2$.

But by II. 6, $AQ . QB = PQ^2 - AP^2$.

$\therefore AQ^2 + QB^2 = 2PQ^2 - 2AP^2 + 4AP^2$

$= 2\{PQ^2 + AP^2\}$.

16. Let the given st. line AB be divided equally at P and unequally at Q.

Then $AQ^2 + QB^2 = 2\{AP^2 + PQ^2\}$ [II. 9].

But $AP^2 = AQ . QB + PQ^2$ [II. 5].

$\therefore AQ^2 + QB^2 = 2\{AQ . QB + 2PQ^2\}$

$= 2AQ . QB + 4PQ^2$.

On II. 11.

17. Let AB be divided in medial section at H.

From A cut off AH′ equal to BH.

Given $AB . BH = AH^2$, and $AH' = BH$.

Now $AB . BH = AH . HB + HB^2$ [II. 3]

$= AH . AH' + AH'^2$;

also $AH^2 = AH . AH' + AH . HH'$ [II. 2].

$\therefore AH . AH' + AH'^2 = AH . AH' + AH . HH'$;

or, $AH'^2 = AH . HH'$.

That is, AH is divided in medial section at H′.

18. If AB is divided in medial section at H,

then $(AH + HB)(AH - HB) = AH^2 - HB^2$ [II. 5]

$= AB . BH - HB^2$ [Hyp.]

$= AH . HB + HB^2 - HB^2$ [II. 3]

$= AH . HB.$

19. Given $AB . BH = AH^2.$

But $AB . BH = XB^2 - XH^2$ [II. 6].

$\therefore\ AH^2 = XB^2 - XH^2,$

or, $AH^2 + XH^2 = XB^2.$

∴ the △, whose sides are AH, XH and XB, is right-angled

20. Given $AB . BH = AH^2.$

But $AB^2 + BH^2 = 2AB . BH + AH^2$ [II. 7]

$= 2AH^2 + AH^2$ [Hyp.]

$= 3AH^2.$

21. Produce DC to K.

Then by II. 6, $AF' . F'C + EC^2 = EF'^2$

$= EB^2$

$= EA^2 + AB^2.$ [I. 47.]

$\therefore\ AF' . F'C = AB^2,$

or, fig. F'K = fig. AD.

To each of these equals add the fig. H'C.

∴ fig. H'D = fig. H'F',

or, $AB . BH' = AH'^2.$

On II. 12 and 13.

22. We have $AB^2 = AC^2 + BC^2 - 2AC . CE.$ [II. 13.]

Also $AC^2 = AB^2 + BC^2 - 2AB . BF.$

By addition $AB^2 + AC^2 = AB^2 + AC^2 + 2BC^2 - 2AB . BF - 2AC . CE,$

or, $AB . BF + AC . CE = BC^2.$

23. Join BD.

Then from the $\triangle$ ADB, since DE is drawn perp. to AB,

$$BD^2 = AD^2 + AB^2 - 2AB \cdot AE \; [\text{II. } 13].$$

Again from the $\triangle$ ADB, since BC is drawn perp. to AD produced,

$$BD^2 = AD^2 + AB^2 - 2AD \cdot AC \; [\text{II. } 13].$$

$$\therefore \; AB \cdot AE = AD \cdot AC.$$

25. Let ABCD be the par^{m}; and let the diag^{s}. meet at O.

Then O is the middle point of AC and BD [Ex. 5, p. 70].

Then by the last Ex., from $\triangle$ ABC, $AB^2 + BC^2 = 2AO^2 + 2OB^2$.

Also, from $\triangle$ CDA, $CD^2 + DA^2 = 2OC^2 + 2OD^2$.

By addition, remembering that AO = OC and OB = OD,

$$AB^2 + BC^2 + CD^2 + DA^2 = 4AO^2 + 4OB^2$$
$$= AC^2 + BD^2 \; [\text{Ex. 1, p. 158}].$$

26. Let ABCD be the quad., and P, Q, R, S, the middle points of the sides AB, BC, CD, DA.

Then PQRS is a par^{m}. [Ex. 9, p. 105]

and AC = 2PQ, also BD = 2SP [Ex. 3, p. 105].

$$\therefore \; AC^2 + BD^2 = 4PQ^2 + 4SP^2$$
$$= 2\{PQ^2 + SR^2 + SP^2 + QR^2\}$$
$$= 2\{PR^2 + SQ^2\} \; [\text{Ex. 25, p. 161}].$$

27. Let ABCD be the rect., and P the given point within it. Let the diagonals meet at O. Then AC = BD [I. 4] and the diag^{s}. bisect one another [Ex. 5, p. 70]. $\therefore$ AO = DO.

Then from $\triangle$ APC, $PA^2 + PC^2 = 2\,[AO^2 + PO^2]$ [Ex. 24, p. 161],

and from $\triangle$ BPD, $PB^2 + PD^2 = 2\,[DO^2 + PO^2]$,

$$\therefore \; PA^2 + PC^2 = PB^2 + PD^2.$$

28. Let ABCD be the quad., and X, Y the middle points of BD, AC. Join YB, YX, YD.

Then from the $\triangle$ ABC, $AB^2 + BC^2 = 2AY^2 + 2BY^2$ [Ex. 24, p. 161];

also from the $\triangle$ ADC, $AD^2 + DC^2 = 2AY^2 + 2DY^2$.

$$\therefore \; AB^2 + BC^2 + AD^2 + DC^2 = 4AY^2 + 2\,(BY^2 + DY^2)$$
$$= AC^2 + 4\,(DX^2 + XY^2) \; [\text{Ex. 24, p. 161}]$$
$$= AC^2 + BD^2 + 4XY^2.$$

29. From the $\triangle APB$, $AP^2 + BP^2 = 2AO^2 + 2OP^2$ [Ex. 24, p. 161].

But both AO and OP are constant,

$\therefore$ $AP^2 + BP^2$ is constant.

30. Let BC be the base, and ABC the $\triangle$ in one of its positions.

Bisect BC at X, and join AX.

Then $BA^2 + AC^2 = 2(BX^2 + AX^2)$ [Ex. 24, p. 161].

But $BA^2 + AC^2$ is constant by hyp.

$\therefore$ $BX^2 + AX^2$ is constant.

And since BX is constant, $\therefore$ AX is constant; and X is a fixed point; $\therefore$ the locus of A is a circle, whose centre is at X.

31. Since CB is a median of the $\triangle ACD$,

$\therefore$ $DC^2 + CA^2 = 2AB^2 + 2BC^2$ [Ex. 24, p. 161].

But $CA^2 = AB^2$ [Hyp.]

$\therefore$ $DC^2 = AB^2 + 2BC^2$.

32. Let ABC be a $\triangle$ rt. angled at B, and let the hypotenuse be divided into three equal parts AE, EF, FC.

Then from the $\triangle ABF$, $AB^2 + BF^2 = 2BE^2 + 2EF^2$ [Ex. 24, p. 161]

and from the $\triangle EBC$, $BE^2 + BC^2 = 2BF^2 + 2EF^2$.

Hence by addition

$$AB^2 + BC^2 + BF^2 + BE^2 = 2BF^2 + 2BE^2 + 4EF^2,$$

or, $AC^2 = BF^2 + BE^2 + 4EF^2$.

But, since $AC = 3EF$, $\therefore$ $AC^2 = 9EF^2$.

$\therefore$ $5EF^2 = BF^2 + BE^2$.

33. Let AX, BY, CZ be the medians of the $\triangle ABC$.

Then $AB^2 + AC^2 = 2AX^2 + 2BX^2$ [Ex. 24, p. 161],

and $AB^2 + BC^2 = 2BY^2 + 2AY^2$,

also $BC^2 + AC^2 = 2CZ^2 + 2AZ^2$.

By addition

$$2AB^2 + 2BC^2 + 2AC^2 = 2(AX^2 + BY^2 + CZ^2) + 2BX^2 + 2AY^2 + 2AZ^2$$

and the doubles of these equals are equal, so that

$$4AB^2 + 4BC^2 + 4AC^2 = 4(AX^2 + BY^2 + CZ^2) + 4BX^2 + 4AY^2 + 4AZ^2$$
$$= 4(AX^2 + BY^2 + CZ^2) + BC^2 + AC^2 + AB^2.$$

Hence $3[AB^2 + BC^2 + AC^2] = 4[AX^2 + BY^2 + CZ^2]$.

34. Let AX, BY, CZ be the medians, intersecting at O.

Then $OA = 2OX$, $OB = 2OY$, $OC = 2OZ$ [Ex. 4, p. 113],

and from the $\triangle BOC$, $OB^2 + OC^2 = 2BX^2 + 2OX^2$.

Again from $\triangle COA$, $OC^2 + OA^2 = 2CY^2 + 2OY^2$,

also from $\triangle AOB$, $OA^2 + OB^2 = 2AZ^2 + 2OZ^2$.

∴ by addition

$$2OA^2 + 2OB^2 + 2OC^2 = 2BX^2 + 2CY^2 + 2AZ^2 + 2OX^2 + 2OY^2 + 2OZ^2,$$

and doubling these equals, we have

$$4OA^2 + 4OB^2 + 4OC^2 = 4BX^2 + 4CY^2 + 4AZ^2 + 4OX^2 + 4OY^2 + 4OZ^2$$
$$= BC^2 + CA^2 + AB^2 + OA^2 + OB^2 + OC^2.$$

Hence $3\{OA^2 + OB^2 + OC^2\} = BC^2 + CA^2 + AB^2$.

35. Let H and K be the middle points of the diags. BD, AC.

Now $PA^2 + PC^2 = 2AK^2 + 2PK^2$ [Ex. 24, p. 161],

and $PB^2 + PD^2 = 2BH^2 + 2PH^2$.

By addition

$$PA^2 + PB^2 + PC^2 + PD^2 = 2AK^2 + 2BH^2 + 4XH^2 + 4XP^2$$
$$= 2AK^2 + 2XK^2 + 2BH^2 + 2XH^2 + 4XP^2$$
$$= XA^2 + XC^2 + XB^2 + XD^2 + 4XP^2.$$

36. Let ABCD be the trapezium, AB and DC being the par[l]. sides, and let DC be less than AB.

Draw CX, DY perp. to AB.

Then $AC^2 = BC^2 + AB^2 - 2AB \cdot BX$ [II. 13],

also $BD^2 = AD^2 + AB^2 - 2AB \cdot AY$;

∴ $AC^2 + BD^2 = BC^2 + AD^2 + 2AB^2 - 2AB \cdot BX - 2AB \cdot AY$.

But $AB^2 = AB \cdot BX + AB \cdot XY + AB \cdot AY$. [II. 1.]

So that $2AB^2 - 2AB \cdot BX - 2AB \cdot AY = 2AB \cdot XY = 2AB \cdot CD$.

Hence $AC^2 + BD^2 = BC^2 + AD^2 + 2AB \cdot CD$.

PROBLEMS.

37. Let H and K be sides of the given sqq., of which H is the greater.

Draw AP equal to H; produce AP to B making PB equal to H; and from PB cut off PQ equal to K.

Then the rect. AQ, QB is that required.

For since AB is divided equally at P and unequally at Q,

$$\therefore \ AQ . QB = PB^2 - PQ^2 \ [\text{II. } 5]$$

$$= \text{sq. on H} - \text{sq. on K}.$$

38. Let BF be the st. line, and K a side of the given sq. [See fig. p. 156.]

On BF describe a semi-⊙, and from any point X in BF, or BF produced, draw XY perp. to BF, making XY equal to K. Through Y draw YHH′ par^l. to BF cutting the semi-⊙ at H and H′. From H (or H′) draw HE perp. to BF.

Then shall $BE . EF = HE^2 = K^2$. [Proof as in II. 14.]

39. Let BE be the side of the rect., and K a side of the given sq. At E draw EH perp. to BE, making EH equal to K. Join BH; and draw HF perp. to BH to meet BE produced at F. Then shall EF be the other side of the rect. For [p. 65, Ex. 4, *or* III. 31] a semicircle described on BF will pass through H.

Hence $BE . EF = EH^2 = K^2$. [Proof as in II. 14.]

40. Let AB and X be the two given st. lines.

Analysis. Let AB be bisected at P and produced to Q.

Then $AQ . QB = PQ^2 - PB^2$. [II. 6.]

Required $AQ . QB = X^2$.

Hence we must have $PQ^2 = X^2 + PB^2$.

Thus the length of PQ may be found by drawing a rt.-angled triangle [I. 47]. And as P is a fixed point, Q is determined.

41. This is the same as dividing a line *externally* in medial section. [See Ex. 21, p. 160, and p. 151.]

42. Draw the rect. ABDH, contained by AB and X.

Produce HA to F, so that

HF . FA = AB . X [II. 14 and Ex. 40, p. 162].

On FA describe a sq. AFGC as in II. 11. Produce GC to meet HD at K.

Then fig. FK = fig. AD.

Take from these equals the common fig. AK.

Then fig. FC = fig. CD.

Or AC^2 = BC . X.

BOOK III.

EXERCISES.

Page 171.

1. The st. line bisecting AB at rt. ∠ˢ is the st. line passing through *both* centres.

2. Let the bisector of ∠ BAC cut BC in D. Then in △ˢ ADB, ADC, it follows that DB = DC and ∠ ADB = ∠ ADC [I. 4]. ∴ AD passes through the centre.

3. If the chords are not parˡ., bisect each at rt. ∠ˢ. If they are parˡ., join their extremities: and bisect each of these new chords at rt. ∠ˢ. The pt. of intersection of the bisectors is the required centre.

4. See fig. to Ex. 1, p. 111. Let A, B, C be the three given pts. Let XO, YO bisect BC, AC at rt. ∠ˢ. Join OB, OC. Then in △ˢ OBX, OCX, OC = OB [I. 4] and in △ˢ OCY, OAY, OC = OA. ∴ OA = OB = OC, and O is the centre of the required ⊙.

5. The required locus is the st. line bisecting at rt. ∠ˢ the st. line joining the two given pts.

6. With the two given pts. as centres, describe circles each of which has a radius equal to the given radius. Again, with the pt. of intersection of these circles as centre, describe a circle with the given radius. This is the circle required.

7. If possible, let a st. line cut the ⊙ in the three points A, E, B, whereof E is between A and B. Then E must fall within the ⊙. But it was assumed to be on the circumference. Hence the st. line AB cannot cut the ⊙ in more than two points.

8. Draw a chord perp. to the st. line joining the given pt. to the centre. This chord will be bisected at the given pt.

9. Let O be the common centre, ABCD the st. line cutting the inner ⊙ in B, C, and the outer ⊙ in A, D. Draw OX perp. to ABCD. Then BX = CX and AX = DX. ∴ difference AB = difference CD.

10. The st. line, through the centre, perp. to one of the par[l]. chords, is perp. to the other [I. 29]. And this st. line bisects both chords. Hence, the st. line joining the middle pts. of two par[l]. chords passes through the centre.

11. The st. line, through the centre, perp. to *one* of the par[l]. chords, is perp. to *all* of them [I. 29]. And this st. line bisects all the chords. Hence it is the required locus.

12. Let the two ⊙s intersect in A, B. Let CAD, EBF be par[l]. st. lines cutting the one circle in C, E and the other in D, F. Then the st. line through the centre of ABEC perp. to the chords AC, BE bisects these chords in P and Q, say. Similarly the st. line through the centre of ABFD perp. to the chords AD, BF bisects these chords in X and Y, say. But PQ is par[l]. to XY [I. 28]. ∴ PX = QY [I. 34]. But CA is double of PA, and AD is double of AX. ∴ CD is double of PX. Similarly EF is double of QY, ∴ CD = EF.

13. Bisect PQ in R, and XY in Z. Let PY, QX intersect in O. Join OR, OZ. Then △PXY = △QXY: and △XOZ = △YOZ. ∴ △PXO = △QYO. Also △POR = △QOR. ∴ △s POR, PXO, XOZ together = △s QOR, QYO, YOZ: i.e. the lines OR, OZ bisect the trapezium PXYQ. But the st. line RZ bisects the trapezium PXYQ [Ex. 8, p. 117, I. 38]. ∴ the st. line RZ coincides with the st. lines OR, OZ: that is, O lies on RZ. Similarly if PX, QY intersect in O′, O′ lies on RZ.

Page 172.

1. The diagonals of a par[m]. bisect one another. ∴ their pt. of intersection is the centre.

2. Let the par^{m}. ABCD be inscribed in a ⊙. Then the diagonals AC, BD intersect in O, the centre of the ⊙ [Ex. 1]. Because AO = BO, ∴ ∠ OAB = ∠ OBA. And because AO = DO,

∴ ∠ OAD = ∠ ODA;

∴ whole ∠ DAB = sum of ∠^{s} ABD, ADB;

∴ ∠ DAB is a rt. ∠ [I. 32].

3. Let C, D be the centres of two ⊙^{s} intersecting in A. Draw AX perp. to CD, and produce it to B, so that BX = AX. Then CA = CB and DA = DB [I. 4]. ∴ B is a pt. on both circles.

Page 182.

1. Let A be the centre of the larger, B of the smaller ⊙. Produce AB to C, making BC = the radius of the smaller ⊙. Then AC is the radius of the larger. ∴ the ⊙^{s} meet at C. Let D be *any* other pt. on the smaller ⊙. Then BD = BC. ∴ AB, BD together = AC. But AB, BD together > AD. ∴ AC > AD. ∴ D cannot be on the circle with centre A and radius AC.

2. Let C be the pt. of contact. Then ABC is a st. line [III. 11]. Because AC = AP, therefore ∠ ACP = ∠ APC. And because BC = BQ, ∴ ∠ BCQ = ∠ BQC. ∴ ∠ APC = ∠ BQC. ∴ AP, BQ are parallel.

Page 183.

1. The required locus is the st. line joining the centre of the given ⊙ to the given pt. [III. 11, 12].

2. Let O be the centre of the given ⊙, OP its radius. On OP take PC equal to the given radius of the circles which are to touch the given ⊙. Then the ⊙ with centre C and radius CP will touch the given ⊙ at P. And OC = the sum or the difference of OP and CP. Hence the required locus is a circle with centre O and radius equal to the sum or the difference of the radius of the given circle and the given radius of the touching circles.

3. Let A, B be the centres of the two circles. From AB cut off AC = radius of ⊙ with centre A. Then BC = radius of ⊙ with centre B. ∴ the ⊙^{s} meet at C. Let D be *any* other pt. on ⊙ with centre B. Then AD, DB together > AB. But BD = BC. ∴ AD > AC. ∴ D cannot be on the circle with centre A and radius AC.

4. Let C be the pt. of contact. Then ACB is a st. line [III. 12]. Because AC = AP, ∴ ∠ ACP = ∠ APC. And because BC = BQ, ∴ ∠ BCQ = ∠ BQC. ∴ ∠ APC = ∠ BQC. ∴ AP, BQ are parallel.

Page 185.

1. Let A, B be the two given pts. Bisect AB in C: draw CX perp. to AB. Then, if CX *coincides with* the given st. line, with *any* pt. X on CX as centre, the circle described with centre C and radius CA will pass through B. But if CX is *par^l. to* the given st. line *no* circle can be described as required. Finally, if CX *cuts* the given st. line in X, the circle described with centre X and radius CA or CB is the circle required.

2. Let A be the given pt., XY the given st. line. Draw AM perp. to XY, and produce it to B, so that BM = AM. All the circles will pass through B. [See Ex. 1, p. 233.]

3. Let A be the centre of the given ⊙, P the given pt. Take PC equal to the given radius, either on PA (produced if necessary) or on AP produced. The circles described with centre C and radius CP will touch the given circle at P: but, if the given radius is equal to the radius of the given circle, one of the two circles so described will coincide with the given circle.

4. Let A be the centre of the given ⊙, B the given pt. Let AB cut the given ⊙ in C and D. The ⊙ described with centre B and radius BC or BD will touch the given ⊙. Hence there are two solutions except when B is on the ○ce of the given ⊙.

5. Let A be the centre of the given ⊙, B the given pt. on it, C the given pt. through which the required ⊙ is to pass. Let the perp. bisector of BC cut AB in O. The ⊙ described with centre O and radius OB is the ⊙ required: but, if C is on the circumference of the given ⊙, O will coincide with A [III. 1], and the ⊙ described will coincide with the given ⊙. The solution is also impossible, if CB is perp. to AB. For then the perp. bisector of BC will not cut AB.

6. Let A and B be the centres of the two given ⊙s X and Y. Describe a circle with centre A and radius equal to the sum or the difference of the radius of the required circle and the radius of X. Describe a circle with centre B and radius equal to the sum or the difference of the radius of the required circle and the radius of Y. Then the circle described with either of the pts. of intersection of these two circles as centre and with the required radius will be the circle required. There are thus in general 8 possible solutions.

7. Let A, B be the middle pts. of the given chords DAF, EBG, so that AD = 3 in. and BE = 4 in. Then, supposing the chords to be on the *same* side of the centre, let AB be produced to C the centre. Then

$$CD^2 = CE^2; \text{ that is, } CA^2 + 9 = CB^2 + 16. \quad \therefore CA^2 - CB^2 = 7.$$

But CA − CB = 1. ∴ CA + CB, that is, AB + 2CB = 7. ∴ CB = 3.

$$\therefore CE^2 = CB^2 + BE^2 = 3^2 + 4^2 = 25. \quad \therefore CE \text{ (the radius)} = 5.$$

8. Let A, B be the centres of the two circles, touching externally at C. Then ACB is a st. line [III. 12]. Draw the parallel diameters DAE, FBG. Then, because AD = AC, ∴ ∠ ADC = ∠ ACD, ∴ ext. ∠ EAC = twice ∠ ACD. Similarly, ∠ CBF = twice ∠ BCG. But ∠ EAC = ∠ CBF, because AE, BF are par[l]. ∴ ∠ ACD = ∠ BCG. ∴ DC, CG are in a st. line; that is, the st. line joining GD passes through C.

9. Let A, B be the centres of the two ⊙[s]; and D, E the pts. where AB cuts the circles. Let PQ be any other st. line cutting the circles in P and Q. Let QA cut the 'A' circle in R. Then QP > QR (III. 8). Also AQ > AE, of which AR = AD; ∴ QR > DE. ∴ *à fortiori*, QP > DE. ∴ DE is less than any st. line PQ cutting the ⊙[s] in P and Q. Similarly, if AB *produced* cut the circles in FG, FG is greater than any of the st. lines PQ cutting the ⊙[s] in P and Q.

10. Let BC be a chord: A the centre, and ADE the radius bisecting BC at rt. ∠[s] in D. Let G be any pt. in BD, and GF perp. to BD, cutting the arc BEC in F. Draw AH perp. to FG. Then AE = AF > HF. And AD = HG; ∴ the whole or remainder DE > whole or remainder GF. In DG take a pt. L, and draw MLK perp. to BC to meet the arc at K. Then $AK^2 = AF^2$, that is, $AM^2 + MK^2 = AH^2 + HF^2$. But $AH^2 > AM^2$.

$$\therefore HF^2 < MK^2, \text{ that is, } GF < LK.$$

11. Let A be any pt. on the circumference: ACB the diameter, AD any other chord. Then CB = CD; ∴ AB = AC + CD > AD. And if AF is nearer to AB than AD, the two sides AC, CD are equal to the two sides AC, CF, but the angle ACF > angle ACD, ∴ base AF > AD [I. 24]. Make ∠ ACE = ∠ ACF. Then AE = AF [I. 4]. But AF > AD. Hence two and only two chords from A can be drawn equal to one another.

Page 187.

1. Since equal chords are equidistant from the centre, the locus is a circle, whose centre is the centre of the given circle and radius is equal to the distance of any of the chords from the centre.

2. Let chords AB, CD cut in E. Let F be the centre. Draw FG, FH perp. to AB, CD. Then in the right-angled △ˢ FEG, FEH, the hypotenuse EF is common and the angles GEF, HEF are equal. ∴ GF = HF [I. 26], ∴ the chords AB, CD are equidistant from the centre, and are therefore equal [III. 14].

3. Take fig. of preceding Ex. Then EG = EH. But BG, HD, the halves of AB, CD [III. 3], are equal. ∴ the whole or remainder BE = whole or remainder CE.

4. With centre, A, on the ○ᶜᵉ of the given ⊙, describe a circle having radius equal to the required *length*, cutting the given circle again in B. From the centre of the given circle, draw a st. line perp. to the required *direction* and equal to the distance of the centre from AB. The chord drawn through the extremity of this st. line, parˡ. to the given direction, will be the chord required.

5. Let O be the centre: and AX, BY, OZ the perpˢ. on PQ. Then OZ = $\frac{1}{2}$ the sum or difference of AX and BY, according as A and B are on the same or opposite sides of PQ [Ex. 18, p. 106]. ∴ the sum or difference of AX and BY = twice OZ = a constant.

Page 189.

1. Let A be the given pt., and B the centre. Draw the chord CAD perp. to AB, and *any* other chord EAF through A. Draw BG perp. to EF. Then the hypotenuse BA > BG. ∴ chord CD < chord EF; that is, CD is the least chord through A.

2. Let XZY, X′Z′Y′ be two chords bisected at Z and Z′ in AB, of which Z is nearer than Z′ to C the middle pt. of AB. Let O be the centre. Then OC, OZ, OZ′ are respectively perp. to AB, XY, X′Y′. ∴ OZ′ > OZ > OC [Ex. 3, p. 101]. ∴ X′Y′ < XY < AB. Hence AB is the greatest length of XY, and XY increases as Z approaches C. When Z coincides with A or B, XY vanishes.

3. Place any chord PQ of required length in the ⊙. [See solution of Ex. 4, p. 187 or IV. 1.] Let O be centre of given ⊙, and AB the given chord upon which the middle pt. of the required chord is to lie. Draw ON perp. to PQ. With centre O and radius ON describe a circle cutting AB in Z and Z′. Then the chord XZY perp. to OZ will be equal to PQ, and be bisected at Z in AB. There is no solution if PQ > AB [Ex. 2]; one solution if PQ = AB; and two solutions if PQ < AB.

Page 195.

1. Draw a diameter (i) at right angles to, (ii) par^{l}. to the given straight line. At either extremity of the diameter draw a line perp. to the diameter. These will be the tangents, (i) and (ii), required.

2. The tangents are perp. to the same diameter, and therefore parallel [I. 28].

3. The pt. of contact is in the line of centres [III. 11, 12], ∴ the st. line drawn from the pt. of contact perp. to the line of centres is a tangent to both circles [III. 16].

4. The radius from the pt. of contact of the inner ⊙ is perp. to the tangent [III. 18], and ∴ bisects the chord of the outer ⊙ [III. 3].

5. The tangents to the inner ⊙ are chords of the outer at equal distances from the centre of the outer; and therefore are equal chords [III. 14].

Pages 196, 197.

1. Let O be the centre of a ⊙ touching AB and AC in B and C. Then OB = OC, OA is common, and ∠^{s} ABO, ACO are rt. ∠^{s} [III. 18], ∴ in the rt.-angled △^{s} AOB, AOC, ∠ OAB = ∠ OAC [Ex. 12, p. 99].

2. Let AO cut BC in D. Then ∠ BAD = ∠ CAD [Ex. 1]. ∴ in △s BAD, CAD, BD = DC and ∠ BDA = ∠ CDA [I. 4].

3. The chords of the outer ⊙ which are tangents to the inner are equal [Ex. 5, p. 195] and are bisected at the pt. of contact [Ex. 4, p. 195]. Hence the tangents, that is the half-chords, are equal.

4. The tangent at an extremity of a diameter is perp. to the diameter. ∴ the chords par^l. to it are bisected by the diameter [III. 3].

5. The required locus is the perp. to the given st. line through the given pt. [III. 19].

6. The required locus is the st. line which is par^l. to the two given st. lines and equidistant from them.

7. The required locus is the pair of bisectors of the angles between the two given st. lines [Ex. 1, p. 196].

8. If the lines are par^l. there is no solution unless the given radius is equal to half the perp. distance between the par^{ls}. If they are not par^l. let them be OX, OY; at O draw OA, OB equal to given radius and perp. to OX, OY respectively. Through A and B draw AP, BP par^l. to OX, OY respectively; then P is the centre of the required ⊙.

9. Let A be given pt. Place a chord CD in the given ⊙ equal to the given st. line. Describe a circle concentric with given ⊙ and with radius equal to the distance of the centre from CD. From A draw a tangent to this ⊙. The tangent is the required chord.

If A is without the circle, the given line must be not greater than the diameter. If A is within the circle, the given line must also be not less than the chord through A perp. to the line joining A to the centre.

10. Let CD, BE be the two par^l. tangents at the extremities of the diameter CAB: and DFE a tangent at F. Then the △s ABE, AFE are identically equal [III. 17 and I. 8]. ∴ AE bisects ∠ BAF. Similarly AD bisects ∠ CAF. ∴ DAE is a rt. ∠ [Ex. 2, p. 35].

11. Let ABCD circumscribe a ⊙ whose centre is O, the pts. of contact of AB, BC, ... being E, F, G, H. ∴ AE = AH. [III. 17. Cor.] Similarly BE = BF, DG = DH and CG = CF. ∴ AE, BE, DG, CG together = AH, DH, BF, CF; i.e. AB, CD together = AD, BC.

12. The opp. sides of a par^m. are equal: and the sum of one pair of opp. sides of a quad^l. circumscribing a ⊙ is equal to the sum of the other pair [Ex. 11]. Hence double of one side = double of the adjacent side. ∴ the circumscribing par^m. is equilateral.

13. Take fig. of Ex. 11. Then, by III. 17. Cor. and I. 8,

∠ AOE = ∠ AOH; ∠ BOE = ∠ BOF;

∠ DOG = ∠ DOH; ∠ COG = ∠ COF.

∴ ∠^{s} AOE, BOE, DOG, COG together = ∠^{s} AOH, DOH, BOF, COF;

i.e. ∠^{s} AOB, COD together = ∠^{s} AOD, BOC.

But these four ∠^{s} together = 4 rt. ∠^{s}.

∴ ∠^{s} AOB, COD = ∠^{s} AOD, BOC = 2 rt. ∠^{s}.

14. Let O be centre. Then OB, being perp. to tangent BD, is par^l. to AD. ∴ ∠ DAB = ∠ ABO. But OA = OB,

∴ ∠ ABO = ∠ OAB.

∴ AB bisects ∠ CAD.

15. Let O be centre, and let AT, BT′ be two equal tangents at A and B. Then OT = OT′ [III. 18 and I. 4]. ∴ locus of T is circle with centre O.

16. See fig. p. 194. Let BCD be given ⊙, EBF the given diameter produced. Draw BA perp. to EB and equal to the required length. Join AE cutting the ⊙ in D. Draw DF perp. to ED meeting EBF in F. Then the tangent DF = AB, the given length. Hence F is the required pt.

17. See fig. p. 194. At E the centre make the angle BEA equal to the complement of half the given angle. Let EA meet the tangent at B in A, and the ⊙ BCD in D. Draw DF perp. to ED meeting EB in F. Then ∠ DFE = ∠ BAE = complement of ∠ BEA = half the given angle. Hence the tangents from F contain the angle required [III. 17. Cor.].

18. Let A be the pt. through which the ⊙ is to pass. B the pt. on the given st. line which the ⊙ is to touch. Join AB. Draw BC perp. to the given line, and make ∠ BAC = ∠ ABC. Then CA = CB. ∴ C is the centre of the required ⊙. [See solution of Ex. 28, p. 238.]

19. Let the line drawn parallel to the 'tangent' line at a distance from it equal to the given radius cut the 'centre' line in O. Then O is the required centre. Two solutions.

20. Describe a ⊙ concentric with the given ⊙, having its radius equal to the sum or difference of the radii of the given ⊙ and of the required ⊙. A pt. of intersection of the ⊙ so described with a line drawn par[l]. to the given line at a distance equal to the radius of the required ⊙, is the centre of the required ⊙. [See solution to Ex. 33, p. 239.]

Page 200.

1. The sum of ∠[s] PAB, PBA is the supplement of the constant ∠ APB [I. 32], and is therefore constant.

2. The ∠[s] QRS, QPS in segment QRPS are equal: and the ∠[s] RQP, RSP in segment RQSP are equal: and the opp. vertical ∠[s] RXQ, SXP are equal.

3. The ∠ PBQ is the supplement of the sum of the ∠[s] BPQ, BQP, i.e. of ∠[s] in the segments BPA, BQA of the two ⊙[s], which are constant.

4. The ∠ PBX = ∠ PAX [III. 21] = vert. opp. ∠ YAQ
= ∠ YBQ [III. 21].

5. The ∠ AOB is the supplement of the sum of the halves of ∠[s] PAB, PBA and is ∴ constant [Ex. 1]. Hence locus of O is the arc of a ⊙ on chord AB. [Converse of Prop. 21, p. 201.]

Page 202.

1. The opp. ∠[s] of a par[m]. are equal; and if a circle can be described about the par[m]., they are together equal to two rt. ∠[s]. ∴ each ∠ is a rt. ∠.

2. The ∠ ABC = ∠ AXY = ∠ AYX = supplement of ∠ XYC. ∴ XBCY is concyclic. [See Converse of Prop. 22, p. 203.]

3. The exterior ∠ = supplement of adjacent interior ∠ = opposite interior ∠. [See solution to Ex. 5, p. 240.]

Page 204.

1. Let ABCD be a quad^l. inscribed in a ⊙. Let BE bisect the int. ∠ at B, and let DE bisect the ext. ∠ at D. Then angle CDE = half the supplement of ∠ADC = half the ∠ABC = ∠CBE. ∴ CBDE is concyclic [Conv. of Prop. 21]. ∴ E is on the ⊙ ABCD.

2. Let ABC be the △, and P, Q, R any points in the ext. arcs BC, CA, AB. Then the sum of the ∠^{s} BAC, BPC = two rt. ∠^{s} [III. 22]. So that the sum of the ∠^{s} BAC, BPC, CBA, CQA, ACB, ARB = 6 rt. ∠^{s}. And of these the ∠^{s} BAC, CBA, ACB = 2 rt. ∠^{s} [I. 32]. ∴ the ∠^{s} BPC, CQA, ARB = 4 rt. ∠^{s}.

3. Let A be the centre, and AB any radius of the ⊙. With B as centre and BA as radius describe a ⊙ cutting the given ⊙ in C and D. CD shall divide the given ⊙ as required. For ∠CBD in the one segment = ∠CAD = twice ∠ in the other segment [III. 20].

Page 205.

1. In fig. III. 23, ∠ACB in smaller segment is greater than ∠ADB in larger segment [I. 16].

2. If P is without the segment, some part of the arc of the segment must lie within the △APB. If Q is any pt. on this part of the arc, ∠AQB > APB [I. 21]. If P is within the segment, AP produced will cut the segment in some pt. Q, so that the ext. ∠APB > the int. opp. ∠ AQB.

3. Let P and X be on BC, Q on CA. Then QX = QC [Ex. 2, p. 108]. ∴ ∠QXC = ∠QCX = ∠PRQ, since PCQR is a par^m. [Ex. 2, p. 104]. ∴ ∠QXP is supplement of ∠PRQ. ∴ P, R, Q, X are concyclic.

4. If Y and Z be the feet of the perp^s. from B and C on CA and AB, P, Q, R, Y and P, Q, R, Z are concyclic. But only one circle can pass through P, Q, R [III. 10]. Hence the six pts. P, Q, R, X, Y, Z are concyclic.

Page 211.

1. Let AB, CD be par^{l}. chords. Then ∠ DAB on arc BD = ∠ ADC on arc AC [I. 29]. ∴ arc BD = arc AC [III. 26].

2. Let AC, BD be equal arcs. Then ∠ ADC on arc AC = ∠ BAD on arc BD [III. 27]. ∴ AB is par^{l}. to CD [I. 27].

3. See fig. III. 26. Let ∠ BGC = ∠ EHF. Then the △^{s} BGC, EHF are identically equal. Also arc BKC = arc ELF, and the whole ○^{ce} KBAC = whole ○^{ce} LEDF, ∴ remainder arc BAC = arc EDF. ∴ ∠ BKC = ∠ ELF [III. 27]. ∴ segments BKC, ELF are similar: and they are on equal chords. ∴ they are equal. [III. 24.] Also △ BGC = △ EHF. ∴ sector BGC = sector EHF.

4. Let chords AC, BD intersect at rt. ∠^{s} in X. Then ∠ AXD = ∠^{s} ABD, BAC together, ∴ the arcs AD, BC subtend at the circumference ∠^{s} together equal to a rt. ∠. And ∠ AXB = ∠^{s} ACB, CBD together, ∴ the arcs AB, CD subtend at the circumference ∠^{s} together equal to a rt. ∠. ∴ arcs AD, BC together = arcs AB, CD = the semicircumference.

5. As in preceding ex., ∠ AXD = sum of the ∠^{s} subtended by AD, BC = ∠ subtended by an arc equal to sum of AD, BC.

6. The ∠ AXB = difference of ∠^{s} ABD, BAC = difference of ∠^{s} subtended at the ○^{ce} by AD, BC = ∠ subtended by arc equal to difference of AD, BC.

7. Let bisector of ∠ APB cut the conjugate arc in Q. Then ∠ APQ = ∠ BPQ, ∴ arc AQ = arc BQ. ∴ Q is the pt. of bisection of the conjugate arc AQB.

8. Let PA, PB cut the other ⊙ in Q and R. Join BQ.

(1) Let Q and R be in PA, PB produced. Then ∠ QBR = sum of ∠^{s} BQA, BPA = sum of ∠^{s} subtended at the ○^{ces} by AB = constant.

(2) Let Q and R be in PA, PB. Then ∠ QBR = difference of ∠^{s} BQA, BPA = difference of ∠^{s} subtended at the ○^{ces} of the two ⊙^{s} by AB = constant.

(3) Let R be in PB and Q in PA produced. Then ∠ QBR = supplement of ∠^{s} BQA, BPA = supplement of ∠^{s} subtended at the ○^{ces} by AB = constant.

9. The ∠ AXY = ∠ ABY = $\frac{1}{2}$B. And ∠ AXZ = ∠ ACZ = $\frac{1}{2}$ C. ∴ ∠ ZXY = $\frac{1}{2}$ (B + C) = complement of $\frac{1}{2}$ A.

10. Let AB, CD be par[l]. chords of a ⊙. Then ∠ ADC = ∠ DAB [I. 29]. ∴ arc AC = arc BD [III. 26]. ∴ chord AC = chord BD [III. 29]. And ∠ CAB = supplement of ∠ ACD [I. 29] = ∠ ABD [III. 22]. ∴ chord BC = chord AD [III. 26, 29].

11. PX and QY subtend at A opp. vertical ∠[s]. Hence arc PX = arc QY. Hence chord PX = chord QY.

12. Each = the common chord [Ex. 1].

13. Since the chord AB is common to the two equal ⊙[s], the arc AB in one = the arc AB in the other [III. 28]; ∴ ∠ APB = ∠ AQB [III. 27]. ∴ BP = BQ.

14. Each of the chords BX, XA, AY, YC subtends an ∠ equal to half the base ∠.

Hence, if the base ∠[s] are each double of the vertical ∠, the pentagon is equilateral.

Page 214.

1. See fig. p. 214. Let tangent at D be par[l]. to AB. Then, if DC be perp. to tangent, the centre is in DC [III. 19]. But DC is also perp. to AB. ∴ DC bisects AB [III. 3.] Hence arc ADB is bisected at D [III. 30].

2. Let CB be the quadrant, A the centre. On AB describe an equilateral △ ADB. Because AD = AB, ∴ D is on the circumference. Bisect ∠ DAB in E. Then the rt. ∠ BAC is trisected by AD, AE [Ex. 6, p. 66]. ∴ the arc BC is trisected at D and E.

Page 217.

1. The ∠ opp. the diameter must be a rt. ∠. Hence the vertex is on the ○[ce]. [Converse of Prop. 21.]

2. The locus is the ⊙ on hyp. as diameter.

3. The locus is a quadrant of the ⊙ whose centre is the pt. of intersection of the rulers, and radius half the length of the rod [III. 31].

4. Each of the ∠[s] PBA, QBA in a semicircle is a rt. ∠. ∴ PB, QB are in a st. line.

5. The line joining the vertex of an isosceles △ to the middle pt. of the base is perp. to the base. Hence the ⊙ on a side as diameter passes through the middle pt. of the base [Ex. 1].

6. Let A be pt. of contact, AB diameter of inner, and ABC of outer ⊙. Draw any chord ADE. Then each of the ∠ˢ ADB, AEC in a semicircle is a rt. ∠. ∴ BD is par[l]. to CE. But B is middle pt. of AC. ∴ D is middle pt. of AE [Ex. 1, p. 104.]

7. Both the circles described on the sides of a △ as diameters must pass through the foot of the perp. from the vertex on the base or base produced.

8. The required locus is the ⊙ whose diameter is the line joining the given pt. to the centre of the given ⊙. If the pt. is *without* the ○ᶜᵉ, the locus is confined within the two tangents to the ⊙. If the pt. is *on* the ○ᶜᵉ, the locus is the ⊙ on the radius through the point as diameter. [See solution of Ex. 40, p. 229.]

9. On the side of the greater of the two given squares as diameter describe a semicircle: from its extremity draw a chord equal to the side of the other given square. The chord completing the △ is the side of the required square [III. 31, I. 47].

10. Let A be a pt. of intersection of two ⊙ˢ; B the centre of one of them. Let the other ⊙ cut the ⊙ described on AB as diameter in C. The chord AC produced will be bisected at C [III. 31, 3].

11. Since the diagonals of a rhombus are at rt. ∠ˢ to one another [Ex. 11, p. 27], ∴ the required locus is the ⊙ described on the given st. line as diameter.

Page 221.

1. If, from one extremity of a chord of a circle, a straight line be drawn making an angle with the chord equal to the angle in the alternate segment, this straight line shall touch the circle.

Take fig. p. 220. Let BA be the diameter through B. Then, if the ∠ DBF is acute, the alternate segment must be > a semicircle. ∴ the diameter BA must fall in this segment.

∴ ∠ DBF = ∠ BAD : add ∠ ABD.

∴ ∠ ABF = ∠ˢ BAD, ABD = a rt. ∠ [I. 32, III. 31].

∴ BF is a tangent. Similarly, if ∠ EBD is an obtuse angle, EB must be a tangent.

2. Each of the ∠ ^{s} made by the tangents with the line joining their pts. of contact is equal to the angle in the alt. segment. ∴ these ∠ ^{s} are equal. ∴ the tangents are equal [I. 6].

3. Let A be pt. of contact, AB, AC the diameters of the given ⊙^{s}. Draw ADE to cut the ⊙^{s} in D and E. Then ∠ ^{s} ADB, AEC in semicircles are rt. ∠ ^{s}. ∴ BD, CE are par^l. and ∠ ^{s} DAB, EAC are in (i) coincident and in (ii) opp. vertical. Hence the remaining ∠ ^{s} ABD, ACE are equal [I. 32]: i. e. the segments DBA, ECA are similar.

4. Draw T′AT the common tangent to the two ⊙^{s} at A. [Ex. 3, p. 195.] Let AX be between AP and AT.

Then ∠ TAX = ∠ APX [III. 32].

And, in (i), ∠ TAX = ∠ AQY :

in (ii), ∠ TAX = ∠ T′AY = ∠ AQY.

∴ in (i) and in (ii) ∠ AQY = ∠ APX.

∴ PX is par^l. to QY.

5. Tangent at A to first ⊙ makes with AO an ∠ equal to OBA in alternate segment. But because O is centre of the other ⊙, OA = OB. ∴ ∠ OBA = ∠ OAB. ∴ AO bisects ∠ between AB and the tangent at A.

6. The tangent at P makes with PAC an ∠ equal to ∠ PBA = ∠ ABD (or its supplement) = ∠ ACD (or its supplement). ∴ tangent at P is par^l. to CD.

7. Let A be pt. of contact, AB chord through A, C the middle pt. of arc cut off by AB; CM, CN perp^s. on the tangent at A and the chord AB. Then ∠ CAM = ∠ ABC [III. 32] and ∠ ABC = ∠ CAB [III. 30]. ∴ ∠ CAM = ∠ CAB. ∴ the ∠ ^{s} CAM, CAN are identically equal [I. 26]. ∴ CM = CN.

Page 223.

1. On the given base describe a segment containing an ∠ = to the given ∠. The pt. or pts. in which the arc of the segment cuts the given st. line give the required vertex.

2. The required vertex is the intersection of the arc of the segment described on the base and containing an ∠ equal to the given ∠, and

(i) The circle, whose centre is an extremity of the base and radius is equal to the given side;

(ii) The st. line parallel to the base at a distance from it equal to the given altitude;

(iii) The circle whose centre is the middle pt. of the base and radius equal to the given median;

(iv) The perp. to the base drawn through the given point.

3. Because arc AP = arc BP; ∴ ∠ ACP = ∠ BCP.

4. Because ∠ ACB = K, and ∠ AXB = $\frac{1}{2}$K,

∴ ∠ XBC = $\frac{1}{2}$K [I. 32] = ∠ AXB.

∴ CB = CX [I. 6]; ∴ AC + CB = AX = required length.

5. On AB, the given base, describe a segment containing an ∠ equal to the given ∠ K; also another segment containing an angle greater by a right ∠ than $\frac{1}{2}$K. From centre A, with radius equal to the given difference of the sides, describe a ⊙ cutting the last drawn segment in X. Join AX and produce it to cut the first segment in C. Then ABC shall be the required triangle.

Let the bisector of ∠ ACB cut BX in D. Then ext. ∠ AXD is greater by the ∠ XDC than ∠ XCD, i.e. than $\frac{1}{2}$K. ∴ ∠ XDC is a rt. ∠. ∴ the △ˢ XCD, BCD are identically equal [I. 26]. Therefore CX = CB. ∴ AX = the difference of AC and CB.

Page 224.

1. Let AB be the base of given segment, produced to C. From A draw AP to meet the arc of the segment at P; join PB, and through C draw CQ par^{l}. to PB to meet AP produced at Q. Then a segment described on the base AC to pass through Q [Ex. 4, p. 171] is that required. For ∠ APB = ∠ AQC [I. 29].

2. Let A be the given point and C the centre of the given ⊙. From A draw the tangent AP; and from P draw PQ, making the ∠ APQ equal to the given angle. From centre C describe a ⊙ to touch PQ; and from A draw a tangent to the ⊙ of construction, cutting the given ⊙ at XY. Then AXY is the required line. For PQ = XY [Ex. 3, p. 197]. Hence the arc PQ = the arc XY [III. 28]; ∴ the angles at the ○ce subtended by these arcs are equal.

Page 227.

1. Bisect AB in G, and CD in H. Draw GF, HF perp. to AB, CD. Then rect. AE, EB + sqq. on EG, GF = sqq. on AG, GF, i.e. rect. AE, EB + sq. on EF = sq. on AF. Similarly, rect. CE, ED + sq. on EF = sq. on CF. ∴ AF = CF. But AF = BF and CF = DF. ∴ A, B, C, D are concyclic. [Or: by *reductio ad absurdum* from Prop. 35.]

2. The shortest chord through a pt. within a ⊙ is the chord bisected at the pt. [Ex. 1, p. 189].

3. On AB as diameter describe a circle. This passes through C [III. 31]. Produce CD to cut the ⊙ in E. Then CE being perp. to the diameter is bisected at D [III. 3]. And rect. AD, DB = rect. CD, DE = sq. on CD.

4. The ⊙ on AB as diameter passes through P and Q [III. 31]. Therefore rect. AO, OP = rect. BO, OQ.

5. Let O be any pt. in AB: and POQ a chord of one circle, ROS a chord of the other. Then rect. PO, OQ = rect. AO, OB = rect. RO, OS. ∴ P, Q, R, S are concylic [Ex. 1].

6. Draw the chord CABD; bisect it in E, and join E to centre F. Then rect. CA, AD + sq. on EA = sq. on EC [II. 5]. And rect. CB, BD + sq. on BE = sq. on EC. But rect. CA, AD = rect. CB, BD. ∴ sq. on EA = sq. on EB. Add sq. on EF. Then sq. on FA = sq. on FB; i.e. A and B are equidistant from centre.

7. Use the lettering of fig. Prop. 35, but take E outside the ⊙. Then rect. EA, EB + sq. on AG = sq. on EG.

Add sq. on GF. Then

rect. EA, EB + sq. on AF = sq. on EF.

Similarly rect. EC, ED + sq. on CF = sq. on EF.

∴ rect. EA, EB = rect. EC, ED.

8. Let AD be diameter of ⊙ ACD, and AE of ⊙ AFE: then ∠^{s} ACD, AFE are rt. ∠^{s} [III. 31], ∴ C and F lie on ⊙ whose diameter is DE. ∴ rect. CA, AE = rect. DA, AF.

Page 229.

1. The sq. on either tangent = rect. contained by the segments of any secant. Hence the tangents are equal.

2. Let AB, the common chord, be produced to C. Then sq. on tangent from C to *either* ⊙ = rect. CA, CB.

3. Let the chord AB produced cut PQ in C. Then sq. on CP = rect. CA, CB = sq. on CQ. ∴ PQ is bisected at C.

4. The sq. on tangent from P to any ⊙ through A and B = rect. PA, PB. Hence sqq. on all the tangents from P are equal.

5. Since ∠^{s} PQB, PCB are rt. ∠^{s}, ∴ Q and C are on the ⊙ whose diameter is PB [III. 31]. ∴ rect. of segments AC, AP = rect. of segments AB, AQ [III. 36].

6. [This is proved in the course of I. 47, p. 90.] *Or:* The ⊙ on BC as diameter passes through D [III. 31]. And AC being perp. to the diameter is tangent at C. ∴ sq. on tangent AC = rect. of segments of secant, AB, AD [III. 36].

Page 232.

1. Let PABQ be a secant cutting the ○^{ce} in A and B. Bisect AB at E. Then EF, the perp. to AB, passes through the centre C [III. 1, Cor.]. Let PABQ move, while A remains fixed, *and* B *approaches* A. Then the perp. EF ultimately coincides with the perp. from A to the tangent at A; and always passes through C.

2. Let AB be the common chord of two ⊙^{s} whose centres are E, F. Then EF bisects AB at right angles in C. When A and B coincide, C must coincide with each; and each becomes the pt. of contact of the two ⊙^{s}. Hence the line joining the centres of two ⊙^{s} which touch one another passes through their point of contact.

3. If the pts. of intersection of two ⊙^{s} come to coincide, the proof that they cannot have the same centre is unaltered.

4. Since two circles cannot cut in 3 points, if two pts. of intersection come to coincide, so that the ⊙ˢ touch, there can be no other point at which they meet.

5. From O the centre draw ON perp. to the given straight line. Then if ON < radius of the ⊙, N is *within* the ⊙, and no st. line can be drawn through N without cutting the *closed* figure in two pts., B and C, say. Now $ON^2 + BN^2 = OB^2 = ON^2 + CN^2$. ∴ as ON increases towards equality to OB or OC, BN and CN decrease towards zero. And ultimately when ON becomes equal to OB, BN and CN become zero: i.e. B and C each coincide with N.

When ON becomes > the radius, any pt. in the straight line is further from the centre than N, ∴ *a fortiori* at a greater distance than the radius.

6. Since the ext. ∠ QAB of the quadrilateral APCB [see fig. p. 232] is equal to the int. and opp. ∠ PCB, ∴ *when* P *coincides with* A and AP becomes the tangent at A, the ext. ∠ made by the tangent with AB becomes equal to the ∠ ACB.

7. Since rect. EA, EB = rect. EC, ED, ∴ when C coincides with D, and EC becomes the tangent at C, rect. EA, EB = sq. on EC.

8. Let AB be the diameter and C any pt. in the circumference. Join CB and produce it to Q. Then ∠ ACQ is a rt. ∠. *When* C *coincides with* B, CQ becomes the tangent at B. Hence AB is perp. to the tangent at B.

9. Let BE bisect the int. ∠ at B, and let DE bisect the ext. ∠ at D. Then E is on the ○ᶜᵉ. When D *coincides with* A, the internal bisector at B meets the bisector of the ∠ between AC and the tangent at A on the ○ᶜᵉ.

THEOREMS AND EXAMPLES ON BOOK III.

I. On the Centre and Chords of a Circle.

Page 233.

2. Let A, B, C be the three given points. Join AB, BC. Bisect AB, BC at rt. angles by st. lines which meet at O. Then O shall be the centre of the required ⊙. [Proof as in III. 25.]

3. Let A, B be the two given points, and PQ the given st. line. Join AB, and bisect it at rt. angles by a st. line which meets PQ at O. Then O is the centre of the required ⊙. [Proof by III. 1. *Cor.*] Impossible when PQ is at rt. angles to AB or AB produced.

4. Let A, B be the given points, and R the given radius. Join AB, and draw CO bisecting it at rt. angles. From centre A (or B) with radius equal to R describe a ⊙ cutting CO at O. Then O is the centre of the required ⊙. [Proof by Ex. 1, p. 233.]

Impossible when the given radius is less than half AB.

5. In the △ˢ ABX, ACY, we have the ∠ABX = the ∠ACY, the ∠AXB = the ∠AYC, and AB = AC; ∴ BX = CY [I. 26].

Or, draw AE perp. to BC; then BE = CE by I. 26, and XE = YE by III. 3. ∴ BX = CY.

6. Let AB be the common chord of two ⊙ˢ whose centres are E, F; and let the st. line parˡ. to AB cut one ⊙ at P, Q and the other at X, Y. Join EF, cutting PQ at O. Then EF is perp. to AB [Ex. 1, p. 171]. ∴ EF is perp. to PQ [I. 29]. ∴ OP = OQ; and OX = OY [III. 3]. Hence PX = QY.

7. Let the two ⊙ˢ intersect at A, B; and let PAQ, XAY be two st. lines equally inclined to AB and terminated by the ○ᶜᵉˢ. Through B draw P′BQ′ parˡ. to PQ. Then P′Q′ = PQ [Ex. 12, p. 171]. Now the ∠XAB = the ∠P′BA; whence it may be shewn XA = P′B, and similarly AY = BQ′. ∴ XY = P′Q′ = PQ.

8. Let the two ⊙ˢ, whose centres are E and F, cut at A, B. Let PAQ be the st. line through A parˡ. to EF and terminated by the ○ᶜᵉˢ, and let XAY be any other st. line terminated by the ○ᶜᵉˢ. Then PQ shall be greater than XY.

By drawing perpˢ. from E, F to PQ it is seen that PQ is double of EF [III. 3]. From E, F draw EG, FH perp. to XY; and from E draw EK perp. to FH. Then XY is double of GH, that is, double of EK. But in the rt.-angled △EKF, EF is greater than EK. ∴ PQ is greater than XY.

9. For, from the two isosceles △ˢ CPA, DAQ, the ∠CPA = the ∠CAP, and the ∠DQA = the ∠DAQ. ∴ the two ∠ˢ XPQ, XQP together = the two ∠ˢ CAP, DAQ. Hence the ∠PXQ = the ∠CAD [I. 32, and I. 13].

10. Let A, B be the points of section of the ⊙s whose centres are C, D. Join CD, and bisect it at O. Join OA, and draw PAQ perp. to OA. Then shall PA = AQ. Draw CE, DF perp. to PQ. Since CE, OA, DF are parl. and CO = OD, ∴ EA = AF [Ex. 13, p. 106]. Hence PA = AQ [III. 3].

11. Let the bisector of the ∠ CPQ meet the ○ce at E. Join CE. Then in the isosceles △ CEP, the ∠ CEP = the ∠ CPE = the ∠ EPQ. ∴ CE is parl. to PQ [I. 27]; that is, CE is perp. to AB. Hence the bisector passes through one or other of the extremities of the diameter at rt. angles to AB.

12. The middle points of the sides of the quadl. (that is, the centres of the ⊙s) are the vertices of a parm. [Ex. 9, p. 105].

Again, the st. line joining the centres of two intersecting ⊙s is perp. to their common chord [Ex. 1, p. 171]. Hence the common chord of two consecutive ⊙s and the common chord of the other two are perp. to parl. lines, and are therefore parl. to one another.

13. Let B, C be the centres of two equal ⊙s which have external contact at A; and let AP, AQ be the two chords at rt. angles to one another. Join PB, QC. Then BC passes through A [III. 12] and the ∠s BAP, CAQ together = one rt. angle; ∴ the four ∠s BAP, BPA, CAQ, CQA together = two rt. angles; ∴ the two ∠s PBA, QCA together = two rt. angles [I. 32]. ∴ PB, QC are parl., and they are also equal; ∴ PQ is equal and parl. to BC.

14. Let A be the given external point, B the centre of the ⊙, Q any point on the ○ce, and P the middle point of AQ. Required to find the locus of P. Bisect AB at O, and join OP. Then because O, P are the middle points of AB and AQ, ∴ OP is half of QB [Ex. 3, p. 105]. That is, OP is of constant length, and O is a fixed point; ∴ the locus of P is a ⊙ whose radius is equal to half the radius of the given ⊙.

15. Complete the ⊙s of which the equal segments are parts. Let C and D be their centres, C and Q being on opp. sides of AB, also P and D. Join CD; then CD will pass through O [Ex. 1, p. 171]. Join CQ, DP. Then in the △s COQ, DOP, we have CO = DO, CQ = DP (for the ⊙s must be equal) and the ∠ COQ = the ∠ DOP.

∴ the △ COQ = the △ DOP identically [Ex. 13, Cor. p. 100]; for the ∠s COQ, DOP are *obtuse* angles.

II. On the Tangent and the Contact of Circles.

Page 235.

1. Let AB be a chord of fixed length, P its middle point, and O the centre of the ⊙. Join OP. Then OP is perp. to AB [III. 3], and is of fixed length for all positions of AB [III. 14]; ∴ the locus of P is a concentric ⊙, which is touched by AB at P, since AB is perp. to the radius OP [III. 16].

2. Let AP, AQ be two tangents drawn from A to a ⊙ whose centre is O, and PR the diameter through P. Then shall the ∠ PAQ be double of ∠ QPR. Join AO, cutting PQ at B. Then AO bisects the ∠ PAQ [III. 17, Cor.], and also bisects PQ at rt. angles [Ex. 2, p. 196]: also PR is perp. to AP [III. 18].

Hence from the rt. angled △s PAO, BPO, the ∠ PAO = the ∠ BPO, each being the comp^{t}. of the ∠ POA. ∴ the ∠ PAQ is double of the ∠ QPR.

3. Let A be the point of contact of the two ⊙s, PAQ the st. line through A terminated by the ○^{ces}, and PB, QC the tangents at P, Q. Then shall PB, QC be par^{l}. Through A draw BAC perp. to the line of centres, meeting PB, QC at B and C. Then BAC touches both ⊙s at A [III. 16]. And because BP = BA [III. 7, Cor.], ∴ the ∠ BPA = the ∠ BAP = the vert. opp. ∠ CAQ = the ∠ CQA (since CA = CQ). That is, the ∠ BPQ = the ∠ CQP; ∴ PB, QC are par^{l}. [I. 27].

4. Let A be a point of intersection of the two ⊙s, PAQ the st. line through A terminated by the ○^{ces}, and PR, QR the tangents at P, Q: and let the tangents at A meet PR, QR at B, C.

Then shall the ∠ PRQ = the ∠ BAC. For, from the two isosceles △s BPA, CAQ [III. 17, Cor.], the ∠ BPA = the ∠ BAP, and the ∠ CQA = the ∠ CAQ: ∴ the two ∠s RPQ, RQP together = the two △s BAP, CAQ. Hence the ∠ PRQ = the ∠ BAC [I. 32, I. 13].

5. Let the two par^{l}. tangents AP, BQ touch the ⊙ at A, B, and cut off the segment PQ from a third tangent whose point of contact is R. Take C the centre. Then shall the ∠ PCQ be a rt. ∠. Join CA, CB, CR.

Then CA, CB being perp. [III. 18] to par^{l}. lines are in one st. line. And PC, QC respectively bisect the ∠s ACR, BCR [III. 17, Cor.]. Hence the ∠ PCQ is half of the four ∠s at C, that is, half of two rt. angles [I. 13].

6. Let O be the centre of the circle; B, C the points of contact of the fixed tangents AP, AQ, and R the point of contact of the third tangent PQ. Then shall the ∠ POQ be constant. Join OB, OR, OC. Now OP, OQ respectively bisect the ∠s BOR, COR [III. 17, Cor.]; ∴ the ∠ POQ is half the reflex ∠ BOC, which is constant, for B, O, C are fixed points.

NOTE. The ∠ POQ = one rt. angle + half the ∠ at A. [See Ex. 36, p. 246.]

7. Let ABCD be the quad^l., and P, Q, R, S the points of contact of the sides AB, BC, CD, DA.

Then AS = AP [III. 17, Cor.] and DS = DR; ∴ by addition AD = AP, DR. Similarly BC = BP, CR. Hence AD and BC together = AP, BP, DR, CR = AB, DC.

8. Let ABCD be a quad^l. in which AB, CD together = BC, DA. By bisecting the two ∠s ABC, BCD describe a ⊙ to touch three sides AB, BC, CD [Ex. 1, p. 196]. Then shall AD also touch this ⊙. For if not, from A draw AD′ touching the ⊙ and cutting CD at D′. Now by hyp., AB, CD together = BC, AD. Also by Ex. 7, AB, CD′ together = BC, AD′. ∴ taking the differences of these equals, DD′ = the difference of AD and AD′; hence either AD′ = AD, DD′, or AD = AD′, DD′; which is impossible [I. 20].

9. Let A be the point of contact, B the centre of the inner ⊙, C of the outer ⊙. Then A, B, C are collinear [III. 11]. Let BC, produced if necessary, cut the inner ○^{ce} at D. Let EF be the chord of the outer ⊙ which touches the inner ⊙ at D, and is therefore perp. to AD [III. 18]; and let PQ be any other chord touching the inner ⊙. From C draw CR perp. to PQ: then R is outside the inner ⊙ [III. Def. 5.] Let CR cut the ○^{ce} at S. Now CS is greater than CD [III. 7]; much more is CR greater than CD; ∴ EF is greater than PQ [III. 15].

10. Let ABC be a △, and F the middle point of the side AB. On BC as diameter describe a ⊙; call its centre D. Join FD and produce it to meet the ○^{ce} at P. Then FP is made up of FD and DP; of which FD is half AC [Ex. 3, p. 105] and DP is half BC: that is FP is half the sum of BC, CA. And a ⊙ described from centre F with radius FP will touch the ⊙ on BC at P; for the centres of the two circles and the point P are collinear. Similarly, the same ⊙ will touch the ⊙ on AC.

11. Let A be the given point, O the centre of the given ⊙, and X the given st. line. In the ⊙ place a chord PQ equal to X [see IV. 1]. With centre O, and radius equal to the perp. from O on PQ, describe a circle, which will be touched by PQ [III. 16]. From A draw ABC to touch the inner ⊙ [III. 17] and to cut the given ⊙ at B, C. Then BC = PQ, being chords at equal distances from the centre of the given ⊙ [III. 18 and 14].

If A is without the ⊙, X must be not greater than the diameter. If A is within the ⊙, X must be not greater than the diameter, and not less than the chord drawn through A perp. to OA.

12. Let O be the given point in the given st. line; and let AB, the given par[l]., cut *any* ⊙ of the system at A, B. Draw AP the tangent at A, and OP perp. to AP. Take C the centre of the ⊙, and join AC, OC: then OC cuts AB in R at rt. angles [Hyp. and I. 29].

Then the ∠ POA = the ∠ ROA, for each is equal to the ∠ OAC [I. 29, I. 5]. Hence △[s] AOP, AOR are identically equal by I. 26. So that OP = OR; and OR is constant, for all ⊙[s] of the system. Now AP is perp. to OP. ∴ AP touches the fixed ⊙ whose centre is O and radius OR.

13. Let A be the centre of the outer, and B of the inner fixed ⊙. Let P be the centre of any third ⊙ touching the first ⊙ at D and the second at E. Then shall AP + BP be constant. Let r_1, r_2, r_3 denote the radii of the three ⊙[s]. Then the points A, P, D and B, E, P are collinear [III. 11 and 12].

And $$AP + BP = r_1 - r_3 + r_2 + r_3 = r_1 + r_2.$$

NOTE. This problem is a special case of the following: If any two circles are touched one internally and one externally by a third circle, the sum or difference of the distances of this third circle from the centres of the given circles is constant.

14. Let PA, PB be any pair of tangents containing the given angle. Take C the centre of the ⊙, and join CA, CP. Now CP bisects the ∠ APB [III. 17, Cor.]. Hence in the △ CAP, the ∠[s] CAP, CPA, and the side CA are constant; ∴ CP is constant [I. 26]. ∴ the locus of P is a concentric ⊙.

15. Let A and B be the centres of the two given ⊙[s], X and Y the given st. lines. At any point C on the O[ce] of the first ⊙ draw a tangent CP equal to X. From A as centre with radius

AP describe a ⊙, and shew that its $\bigcirc^{ce}$ is the locus of points from which tangents of the required length may be drawn to first given circle. Proceeding in a similar manner with the second circle, we see that the points common to the two locus-circles satisfy the conditions. There are two solutions, one solution or no solution according as the loci-circles intersect, touch one another, or do not meet.

16. *Lemma.* If ABC is a triangle, and X a point in the base, such that $AB^2 \sim AC^2 = BX^2 \sim CX^2$, then AX is perp. to BC. This is the converse of Ex. 7, p. 92, and may be proved indirectly from that theorem.

Let A, B, C be the centres of the three ⊙^{s}. Then BC, CA, AB pass respectively through P, Q, R the points of contact [III. 12]. Let the common tangents at Q and R meet at O. *Join* OP. Then OP shall touch the ⊙^{s} (B) and (C) at P. Join OA, OB, OC.

Now $OB^2 = OR^2 + BR^2$, and $OC^2 = OQ^2 + CQ^2$ [III. 18, I. 47]

Hence by subtraction, remembering that OQ=OR [III. 17, Cor.]

$$OB^2 \sim OC^2 = BR^2 \sim CQ^2 = BP^2 \sim CP^2.$$

∴ OP is perp. to BC. (Lemma.)

∴ OP touches the ⊙^{s} (B) and (C).

∴ OP = OQ [III. 17, Cor.] = OR.

Common Tangents. Page 236.

17. (i) The two direct tangents only can be drawn in this case: for when we attempt to draw the transverse tangents we find the point B *within* the circle of construction.

∴ no tangent can be drawn to it from B.

(ii) Here the two direct tangents may be drawn, and the two transverse tangents become *coincident*. For B will fall on the $\bigcirc^{ce}$ of the circle of construction; hence only one tangent (or two coincident tangents) may be drawn to it from B.

(iii) Hence for similar reasons the two direct tangents are *coincident* and the two transverse tangents are *impossible.*

(iv) Both direct and transverse tangents are *impossible.*

18. In this case the ⊙ of constr. is reduced to a point. Proceed thus :—join AB the centres of the given ⊙ˢ, and draw AD, BE perp. to AB, cutting the ○ᶜᵉˢ in D and E. Join DE, which will be one direct common tangent. [Proof by I. 28, 33, 34, and III. 16.]

19. Let a pair of common tangents touch the greater ⊙ at D, D′, the smaller at E, E′, and cut one another at P.

Then by III. 17, Cor., PD = PD′, and PE = PE′.

∴ for direct tangents PD − PE = PD′ − PE′;

and for transverse tangents PD + PE = PD′ + PE′;

∴ in either case DE = D′E′.

If the ⊙ˢ are equal, then the direct common tangents are equal [I. 34]. Or again, with the fig. of p. 236, DE = BC; similarly D′E′ = BC′; but BC = BC′, ∴ DE = D′E′.

20. Let the direct common tangents DE, D′E′ touch the ⊙ˢ whose centres are A, B at D, E and D′, E′, and cut one another at P. Join PB, BE, BE′. Then in the △ˢ PEB, PE′B, we have BE = BE′ and BP common, also the ∠ˢ PEB, PE′B are rt. ∠ˢ [III. 18];

∴ ∠ EPB = ∠ E′PB [Ex. 12, p. 99].

That is, the centre B lies on the bisector of the ∠ between the common tangents. Similarly the centre A lies on the same bisector. Therefore the points A, B, P are collinear.

21. Let B, C be the centres of the two given ⊙ˢ: then BC passes through A [III. 12]. Join BP, CQ.

Then the sum of ∠ˢ BAP, CAQ = the sum of ∠ˢ BPA, CQA
= the sum of the compᵗˢ. of ∠ˢ APQ, AQC [III. 8]
= ∠ PAQ. [I. 32.]

Hence ∠ PAQ is half of two rt. ∠ˢ; that is, the ∠ PAQ is a rt. ∠.

22. Let B, C be the centres of the two given ⊙ˢ; then BC passes through A [III. 12]. At A draw the common tangent to meet PQ at X. Then XA = XP and XA = XQ [III. 17, Cor.].

∴ a ⊙ described on PQ as diameter passes through A, and touches BC, for XA is perp. to BC [III. 16].

23. Let the bisector of the ∠ PCA meet PQ at R. Join RA. Then by I. 4, the △s CPR, CAR are identically equal; ∴ ∠ RAC is a rt. ∠; hence RA is the tangent to both ⊙s at A [III. 16]. Thus the bisector of the ∠ PCA meets PQ at the point at which it is cut by the tangent at A. Similarly the bisector of the ∠ QC′A meets PQ at the same point: that is, the bisectors intersect on PQ; and are at rt. angles, for they are also the bisectors of the ∠s PRA, QRA [Ex. 2, p. 35].

24. Let C, C′ be the centres of the two ⊙s. From centre C with radius equal to the difference of the radii of the given ⊙s, describe a ⊙ to cut CC′ at X, Y; and from C′ draw the tangent C′P′. Then [Ex. 17, p. 236] PQ = C′P′.

∴ the sq. on PQ = the sq. on P′C′

= the rect. C′X, C′Y

= 2C′A . 2CA

= the rect. contained by the diameters.

25. Let A be the centre of the ⊙ to which the tangent is to be drawn, and B the centre of the ⊙ which is to cut off from the tangent an intercept equal to K. In the ⊙ (B) place a chord equal to K, and describe a concentric ⊙ to touch this chord (i.e. to pass through its middle point). Then draw a common tangent to the ⊙ (A) and the ⊙ of construction. Then the ⊙ (B) will cut off from this tangent a part equal to K [Ex. 5, p. 195].

Impossible when K is greater than the diameter of the ⊙ (B), or when, of the circle (A) and the ⊙ of construction, one falls within the other. In general there are four solutions.

26. Let A and B be the centres of the given ⊙s, H and K the two given lines. Place chords equal to H and K respectively in the ⊙s (A) and (B), and describe concentric ⊙s touching these chords. Then draw a common tangent to the two ⊙s of construction. From this tangent the two given ⊙s will cut off parts equal to H and K [Ex. 5, p. 195].

PROBLEMS ON TANGENCY. Page 238.

Loci. (i) The st. line which bisects the line joining the given points at rt. angles.

(ii) The st. line perp. to the given st. line at the given point.

(iii) The radius through the given point, indefinitely produced both ways.

(iv) Two st. lines par^{l}. to the given line, one on each side of it, at a perp. distance from it equal to the radius of the touching circles.

(v) Two concentric circles, whose radii are $r_1 + r_2$ and $r_1 \sim r_2$, where r_1 is the radius of the given circle, and r_2 the radius of the circles which touch it externally or internally.

(vi) The two st. lines which bisect internally and externally the angle between the two given st. lines.

27. The three given st. lines are supposed to be of infinite length. The locus of the centres of ⊙^{s} touching any pair must be the internal and external bisectors of the angle between them. Four *different* centres will be given by the intersection of these loci, corresponding to what are known as the *inscribed* and *escribed* ⊙^{s} of the △ formed by the three given lines.

28. Let AB be the given st. line, C the given point in it, and D the other point through which the required ⊙ is to pass. Then since the required ⊙ is to touch AB at C, its centre must lie on the st. line through C perp. to AB.

Again, since the required ⊙ is to pass both through C and D, its centre lies on the st. line which bisects CD at rt. angles. Therefore O, the intersection of these loci, is the centre of the required ⊙.

One solution: except when D is in AB, then impossible, for the loci will in that case never meet.

29. Let C be the centre of the given ⊙, A the point on its ○^{ce}, and D the other point through which the required ⊙ is to pass.

Then since the required ⊙ is to touch the given ⊙ at A, ∴ its centre must lie on CA, or CA produced [III. 11, 12].

pairs of coincident solutions, two distinct, and the remaining two impossible; and if x is less than r_1, six solutions are possible.

Finally, all eight solutions are possible, if

$$r_1 > 2r \quad \text{and} \quad x < r_1 - 2r.$$

34. Let AB be the given st. line, and r_1, r_2 the radii of the given ⊙s.

Describe a ⊙ of radius r_1 to touch AB.

(i) Then the centre of 2nd required circle must lie on one or other of the concentric ⊙s whose radii are $r_1 + r_2$ or $r_1 \sim r_2$.

(ii) The centre of the 2nd required ⊙ must also lie on the st. line par^l. to AB at a distance from it equal to r_2, and on the same side of it as the 1st ⊙.

Hence theoretically we have four solutions.

The ⊙, whose radius is $r_1 + r_2$, will always give two possible distinct solutions. The ⊙ whose radius is $r_1 \sim r_2$ gives two coincident solutions.

35. Let PQ be the given line, and C the centre of the given ⊙; and let a second ⊙, whose centre is F, touch the given ⊙ at E and PQ at A. Then shall AE produced meet the ○^{ce} of the given ⊙ at D, an extremity of the diameter perp. to PQ. Join DC, FA, CF. Then CF passes through E [III. 12].

Now ∠ FAE = ∠ FEA, because FA = FE;
= ∠ CED [I. 15]
= ∠ CDE, because CD = CE.

∴ DC is par^l. to FA; but FA is perp. to PQ [III. 18].

∴ AE passes through an extremity of the diameter perp. to QP.

36. Because CD = CE, ∴ ∠ CDE = ∠ CED;
= ∠ FEA [I. 15].

Also ∠ CDE = alt. ∠ FAE [I. 29].

∴ ∠ FEA = ∠ FAE; ∴ FE = FA.

Now FE produced passes through the centre C, and FA is perp. to PQ;

∴ a ⊙ described from centre F with radius FA satisfies the required conditions [III. 12 and 16].

(i) If PQ is without the given ⊙, then the ⊙ derived from AD has external contact, that derived from AB internal contact (the given ⊙ being within the other).

(ii) If PQ touches the given ⊙, then the ⊙ derived from AD has external contact, that from AB is impossible.

(iii) If PQ cuts the given ⊙, then both ⊙ˢ touch externally, or both internally, according as the point A is without or within the given ⊙.

37. Let PQ be the given st. line, and E the given point on the ⊙ of which C is the centre. Draw the diameter BD perp. to PQ. Join DE (or BE), and produce it to meet PQ at A. Draw AF perp. to PQ; and join CE, producing it to cut AF at F. Then F shall be the centre of the required ⊙. [Proof as in Ex. 36.]

38. Let BD be given st. line, and D the given point in it. Let F be the centre of the given circle. [See fig. p. 239.]

To the given ⊙ draw a tangent AP perp. to BD, A being the point of contact. Join AD, meeting the ○ᶜᵉ at E. Join FE and produce it to meet BD in C.

Then C shall be the centre of the required circle. [Proof as in Ex. 36.] Two solutions, since two tangents may be drawn to the given ⊙ perp. to BD.

Orthogonal Circles. Page 240.

39. Let A and B be the centres of the two ⊙ˢ, and C, C′ their intersections. Then ∠ ACB = ∠ AC′B [I. 8].

And the angles between the tangents at C, and the tangents at C′ are respectively supplementary to the ∠ˢ ACB, AC′B.

40. This follows immediately from III. 19.

41. This follows from Ex. 40, by the aid of I. 47.

42. It follows from Ex. 40 that the required locus is the tangent to the given circle at the given point.

Again, since the required ⊙ is to pass through the points A and D, its centre must lie on the st. line which bisects AD at rt. angles. ∴ O, the intersection of these loci, is the centre of the required ⊙.

One solution: except when D lies on the tangent at A; then impossible, for the loci in that case will never meet.

30. Let r be the given radius, AB the given st. line, C the given point.

(i) Then since the required ⊙ is to touch AB, its centre must lie on one or other of the two st. lines par[l]. to AB and at a distance from it equal to r.

(ii) Again, since the required ⊙ is to pass through C, its centre must lie on the ○[ce] of a ⊙ of which C is the centre, and r the radius.

Hence the intersections of either st. line in (i) with the ⊙ in (ii) will give centres of the required ⊙. *Theoretically* there will be four solutions.

(i) If C is in AB, the circle-locus will *touch* both of the par[ls]., and there will be two pairs of coincident solutions.

(ii) If C is not in AB, the circle-locus can only cut that parallel which is on the same side of AB as C: thus of the four theoretical solutions, two will be impossible, and the other two will be distinct, coincident or impossible as the distance of C from AB is less, equal to, or greater than $2r$.

31. Let A and B be the centres of the given ⊙[s], and r_1, r_2 their radii; and let r be the radius of the required circle.

(i) Then the centres of all ⊙[s] of radius r which touch the ⊙ (A), lie on one or other of the concentric ⊙[s] whose radii are $r_1 + r$, or $r_1 \sim r$ respectively.

(ii) Again, the centres of all ⊙[s] of radius r which touch the ⊙ (B), lie on one or other of the concentric ⊙[s] whose radii are $r_2 + r$ or $r_2 \sim r$.

Hence the intersections of either ⊙ in (i) with either ⊙ in (ii) give centres of the required ⊙.

Thus theoretically we get *eight* solutions. Which of them are real, and which impossible will be found to depend upon the relative magnitudes of r_1, r_2, and r; and also upon the relative position of the two given ⊙s—whether one is without the other, one within the other, or whether they intersect.

32. Let AB, CD be the two given st. lines, and r the radius of the req. ⊙.

(i) Then all ⊙s of radius r which touch AB must have their centres on one or other of the st. lines par[l]. to AB, and at a perp. distance from it equal to r.

(ii) Similarly all ⊙s of radius r which touch BC must have their centres on one or other of the st. lines par[l]. to BC, and at a perp. distance from it equal to r.

Hence the intersections of either st. line in (i) with either st. line in (ii) gives a centre of the required ⊙.

Thus there will be *four* solutions, all of which will be real, when the given lines intersect. If AB and CD are par[l]., the method fails.

In this case there will be no real solution, unless r = half the perp. distance between AB and CD: then there will be an infinite number of solutions.

33. Let AB be the given st. line, r_1 the radius of the given ⊙, r the radius of the required ⊙.

(i) Then the centres of all ⊙s of radius r, which touch the given ⊙, will lie on a concentric ⊙ of radius $r_1 + r$ or $r_1 \sim r$.

(ii) And the centres of all ⊙s of radius r, which touch AB, will lie on one or other of the st. lines par[l]. to AB at a distance from it equal to r.

Hence the intersections of either ⊙ in (i) with either st. line in (ii) give centres of the required ⊙.

Thus theoretically there are eight solutions.

Suppose r_1 greater than r.

Let x denote the distance of AB from the given centre. Then if x is greater than $r_1 + 2r$ all the solutions are impossible.

If $x = r_1 + 2r$ then two solutions are coincident, the rest impossible.

If x lie between r_1 and $r_1 + 2r$, two solutions are real (and distinct), the rest impossible. Again, if $x = r_1$, there are two

43. Let A be the centre of the given ⊙, P the point on its ○^{ce}, and Q another point.

Draw PR the tangent at P. Then the centre of the required ⊙ must lie on this tangent [Ex. 40]. Again, the centre of the required ⊙ must lie on the line which bisects PQ at rt. $\angle^s$. Hence the centre is determined.

III. On Angles in Segments, and Angles at the Centres and Circumferences of Circles.

Page 240.

2. Let the chords AB, CD intersect without the ⊙ at E. Join AD.

Then $\angle$ AEC = $\angle$ ADC ~ $\angle$ DAB. I. 32.

That is, $\angle$ AEC = the difference of the $\angle^s$ at the ○^{ce} subtended by the arcs AC, BD; or the $\angle$ at the centre subtended by half the difference of the arcs AC, BD.

3. Let AB, CD two chords of a ⊙ intersect at rt. $\angle^s$ at E.

Then by Ex. 1, the $\angle$ AED is equal to the sum of the $\angle^s$ subtended at the ○^{ce} by AC, BD.

That is the sum of the arcs AC, BD subtend a rt. angle at the ○^{ce}; or, the sum of the arcs is equal to a semi-circumference [III. 31. Converse].

4. For the $\angle$ AXY = the $\angle$ subtended at the ○^{ce} by the sum or diff. of the arcs AQ, PB. [Ex. 1, p. 240.]

Similarly the $\angle$ AYX = the $\angle$ subtended at the ○^{ce} by the sum or diff. of the arcs QC, AP.

But by hyp. the arcs AQ, PB = the arcs QC, AP respectively.

∴ $\angle$ AXY = $\angle$ AYX ; ∴ AX = AY.

5. Let ABCD be a quad^l. inscribed in a ⊙, having one side DA produced to E.

Then the $\angle^s$ DAB, DCB together = two rt. angles [III. 22],
and the $\angle^s$ DAB, BAE together = two rt. angles [I. 13].

Hence $\angle$ BAE = $\angle$ DCB.

6. Let the two ⊙s intersect at A, B, and let PAQ, XBY be the two st. lines terminated at the ○ces. Join AB. [In the figure taken A lies between P and Q, and B between X and Y.]

Then the ∠s XPA, XBA together = two rt. angles [III. 22], and the ∠ ext. XBA = the ∠ AQY. [Ex. 5.]

∴ the ∠s XPA, AQY together = two rt. angles.

∴ PX and QY are parl. [I. 28].

7. Join PR, QR. Then PR, QR shall be in one st. line.

For ∠ PRB = the suppt. of ∠ PCB [III. 22]

= the suppt. of ∠ BAD [Ex. 5.]

= the suppt. of ∠ BRQ.

∴ P, R, Q are collinear [I. 14].

8. Let ABC be the △, rt.-angled at B, and let the ⊙ on AB as diameter meet AC at D. Then the tangent at D shall bisect BC at E. Join BD.

Since ABC is a rt. ∠, BC is the tangent at B [III. 16]

∴ BE = DE. [III. 17, Cor.]

And since BDC is a rt. angle [III. 31], it follows that

∠ EDC = ∠ ECD. ∴ DE = EC.

Hence BE = EC.

9. Let A, B, C be the three points. Through B draw any st. line BX, in which take *any* point P on the same side of BC as A. At P in BP make ∠ BPQ equal to ∠ BAC. Through C draw CD parl. to PQ. Then D is a point on the ⊙.

For ∠ BDC = ∠ BPQ [I. 29] = ∠ BAC [constr.].

Hence the points B, A, D, C are concyclic [III. 21, Cor.].

10. Let A, B, C be the given points. On the side of CB remote from A make ∠ CBD equal to ∠ BAC.

Then BD is the tangent at B [III. 32, Converse].

11. Let E be the centre of the second ⊙. Join AB, EB, DE and EC. [In the fig. taken ACD lies between E and B.]

Then $\angle$ CEB = $\angle$ CAB [III. 21].

And $\angle$ DEB is double of $\angle$ DAB [III. 20].

∴ $\angle$ DEB is also double of $\angle$ CEB,

∴ $\angle$ DEC = $\angle$ BEC.

Hence $\triangle^s$ DCE, BCE are identically equal [I. 4].

12. Join BQ. Then $\angle$ APB = $\angle$ PQB [III. 32].

But $\angle$ PQB = $\angle$ BPQ [III. 27],

∴ $\angle$ APB = $\angle$ BPQ.

13. For $\angle$ BAD = $\angle$ ACB [III. 32],

and $\angle$ BAC = $\angle$ ADB [III. 32].

∴ $\angle$ ABC = $\angle$ DBA [I. 32].

14. Let AB be the chord, C *any* point on the exterior segment.

Let AC, BQ meet interior segment at P and Q. Then shall PQ be constant. Join AQ.

Then $\angle$ AQB = sum of $\angle^s$ ACQ, CAQ [I. 32].

∴ $\angle$ CAQ = diff. of $\angle^s$ AQB, ACB, both of which are of constant magnitude [III. 21].

∴ $\angle$ CAQ, i.e. the $\angle$ PAQ, is constant.

Hence the arc PQ is constant [III. 26].

15. If all the given $\triangle^s$ stand on a fixed base BC, and have a given vertical angle, they also have the *same circumscribed circle* [III. 21. Converse].

Take BAC, any one of these $\triangle^s$, and let the bisector of the $\angle$ A meet the circum-circle at X.

Then since $\angle$ BAX = $\angle$ CAX (hyp.), ∴ arc BX = CX [III. 26].

∴ X, being the middle point of the arc BC, is same for all triangles of the series.

16. Draw CF perp. to AE. Then AE bisects the ∠ BAC [III. 27]. Hence ∠ FCB = half the diff. of the ∠s at B and C [Ex. 7, p. 109]. Now DE, EA are respectively perp. to BC, AE.

∴ ∠ DEA = ∠ BCF [Ex. 3, p. 65]

= half the diff. of the ∠s at B and C.

17. Let BC be the chord of the ext. ⊙, and D its point of contact with the int. ⊙. Then shall AD bisect ∠ BAC.

At A draw the common tangent AT.

Then ∠ DAC = ∠ DAT − ∠ CAT

= ∠ ADC − ∠ ABD [III. 17, Cor., III. 32]

= ∠ BAD [I. 32].

18. Let BC, the chord of the ext. ⊙, cut the int. ⊙ at P, Q. Let A be the point of contact of the two ⊙s.

Then shall ∠ BAP = ∠ CAQ.

At A, draw the common tangent AT.

Then ∠ BAP = ∠ TAP − ∠ TAB

= ∠ AQP − ∠ ACB [III. 32]

= ∠ QAC [I. 32].

On the Orthocentre of a Triangle. Page 244.

In an acute-angled △ the orthocentre is within the △.

In an obtuse-angled △ the orthocentre is without the △.

22. For, in the fig. of p. 243, produce ED to X.

It has been shewn that ∠ EDC = ∠ FDB [Ex. 20, p. 239].

But ∠ EDC = ∠ BDX [I. 15]; ∴ ∠ FDB = ∠ BDX.

That is, the ext. ∠ FDX is bisected by BD: and so on for the other ∠s of the pedal △.

The latter part of the proposition may be solved in a similar manner.

23. For, with the fig. of p. 227, since the ∠s AFO, AEO are rt. angles (hyp.), ∴ the four points A, F, O, E are concyclic.

∴ the ∠s FAE, FOE together = two rt. angles [III. 22].

That is, the ∠s BAC, BOC together = two rt. angles [I. 15].

24. For, with the fig. of p. **243**, consider the △OBC.

Here BF is the perp. from B on the opp. side CO produced:
and CE is the perp. from C on the opp. side BO produced.

Now BF and CE intersect in A, and AO produced is perp. to BC. Hence A is the orthocentre of the △OBC.

25. Consider the ⊙s circumscribed about the △s ABC, OBC; and let X be any point on the ○ce of the ⊙ BOC, on the side of BC remote from O.

Then the ∠s BOC, BXC are supplementary [III. 22],
and the ∠s BOC, BAC are supplementary [Ex. 23, p. 244];

∴ ∠ BXC = ∠ BAC.

Hence the segments BAC, BXC are equal, for they stand on equal bases, and contain equal angles [III. 24], ∴ the circles of which these segments are parts are equal.

26. Consider the △ FAB. BD is perp. to the side AF [III. 31], and AE is perp. to BF for the same reason:

∴ G, their point of intersection, is the orthocentre of the △AFB.

∴ FG (produced, if necessary) is perp. to AB [Ex. 19, p. 242].

27. It will be seen that D is the orthocentre of the △EAC.

For AD, being parl. to BC, would meet EC at rt. angles [I. 29].

And CD, being parl. to AB, would meet EA at rt. angles.

Hence ED, produced if necessary, must meet AC at rt. angles [Ex. 19, p. 242].

28. For ∠ BCK = ∠ BAK, in same segment
= compt. of ∠ AKB [III. 31]
= compt. of ∠ ACB [III. 21]
= ∠ OBC [p. 243, Ex. 20].

Similarly ∠ KBC = ∠ BCO;

∴ BO is parl. to KC, and BK parl. to OC [I. 27].

29. For, with the figure of the last exercise, since BOCK is a par^{m}., ∴ the diagonals bisect one another [Ex. 5, p. 70]. That is, KO passes through the middle point of BC. Hence the st. line joining O to the middle point of BC, passes through K.

30. For, from Ex. 29, we see that the st. line joining the orthocentre to the middle point of the base passes through an extremity of the diam^{r}. drawn from A.

∴ ∠ APQ is a rt. angle [III. 31]; and since AP is also perp. to BC, ∴ PQ is par^{l}. to BC [I. 28].

31. Let SX be the perp. drawn from S the centre of the circum-⊙ on BC. Then by [Ex. 29, p. 245] AS and OX meet the ○^{ce} at the same point Q. And SX, passing through the middle point of AQ, is par^{l}. to AO; ∴ SX is half of AO [Ex. 3, p. 105].

32. Let S be the centre of the ⊙ circumscribed about the △ABC, and A′, B′, C′ the centres of the ⊙^{s} about the △^{s} OBC, OCA, OAB.

Then it follows from [Ex. 25, p. 244] that SA′ and BC bisect one another at rt. angles. Also SB′ and AC.

Hence by [Ex. 31, p. 245] AO = A′S Similarly OB = SB′.

Again SA′ and AO are par^{l}., for both are perp. to BC.

Similarly SB′ and BO are par^{l}. ∴ ∠ AOB = ∠ A′SB′.

∴ A′B′ = AB. [I. 4] Similarly B′C′ = BC and C′A′ = CA.

It may be noticed that in the △^{s} ABC, A′B′C′ the orthocentre of each is the circumcentre of the other.

33. Let AP meet RQ in X. Consider the △PRX.

The ∠ XPR = ∠ ACR [III. 21] $= \frac{C}{2}$.

The ∠ PRX = ∠ PRC + ∠ CRQ

= ∠ PAC + ∠ CBQ [III. 21] $= \frac{A}{2} + \frac{B}{2}$.

∴ the ∠^{s} XPR, PRX together $= \frac{A}{2} + \frac{B}{2} + \frac{C}{2}$ = one rt. angle.

∴ AP is perp. to RQ.

34. Let A be the vertex, O the orthocentre, and S the centre of the circum-⊙.

From centre S with radius SA describe a ⊙.

Join AO and produce it to meet the $\bigcirc^{ce}$ at G.

Bisect OG at D, and draw the chord BC perp. to AG. Join AB, AC. Then ABC shall be the required △. Proof follows from [Ex. 21, p. 244].

Loci. Page 247.

38. Let BC be the given base, and BAC any △ of the system, having the vertical ∠ BAC constant in magnitude, but not fixed in position. Let the bisectors of the exterior angles at B and C intersect at I_1.

Then ∠ CBI_1 is half the supplement of the ∠ B.

That is, ∠ CBI_1 is the complement of the ∠ $\frac{B}{2}$.

And ∠ BCI_1 is the complement of the ∠ $\frac{C}{2}$.

But in △ I_1BC

$$\angle I_1 + \angle I_1BC + \angle I_1CB = \text{two rt. angles [I. 32].}$$

Hence $\angle I_1 = \frac{B}{2} + \frac{C}{2}$

$= \text{comp}^t.\text{ of } \angle \frac{A}{2}$, and this is constant.

∴ since the base BC is fixed, the locus of I_1 is the arc of a segment of a circle [III. 21, Cor.].

Note. The locus of I in Ex. 36 and the locus of I_1 are conjugate arcs of the same ⊙.

39. Let the bisectors meet at X.

Then ∠ PAB, QBA together = two rt. angles [I. 29].

∴ ∠ XAB, XBA together = one rt. angle [Hyp.].

∴ ∠ AXB is a rt. angle [I. 32].

And since AB is fixed, the locus of X is a circle on AB as diameter [III. 31. Converse].

40. Let A be the fixed point, C the centre of the ⊙, and APQ any chord through A, meeting the ○ᶜᵉ at P, Q. Let X be the middle point of PQ. Then CX is perp. to PQ [III. 3].

That is, the ∠ AXC is a rt. angle, and since AC is a fixed base, the point X lies on the ○ᶜᵉ of a ⊙ on AC as diam.

(i) If A is external, the locus is that part of the ⊙ on AC which is intercepted within the given ⊙.

(ii) If A is on the ○ᶜᵉ, the locus is a complete ⊙ described on the radius AC as diam., and having internal contact with the given ⊙.

(iii) If A is internal, the locus is a complete ⊙ falling within the given ⊙.

41. Let A be the given point, and B the common centre of the concentric ⊙ˢ. Let P be the point of contact of a tangent from A on any one of these ⊙ˢ. Then APB is a rt. angle [III. 18].

And since A and B are fixed points, the locus is a circle on AB as diam.

42. Let A, B be the fixed points on the ○ᶜᵉ, PQ the arc of constant length but variable position. Let AP, BQ intersect at X. To find the locus of X. [In the fig. taken AP, BQ intersect without the ⊙]. Join PB.

Then ∠ APB = sum of ∠ˢ AXB, PBX [I. 32],

or ∠ X = diff. of ∠ˢ APB, PBQ.

But these are constant angles, being subtended by the constant arcs AB and PQ [III. 21]; ∴ the ∠ X is constant.

∴ the locus is the arc of a segment described on AB [III. 21, Cor.]. When AP, BQ intersect within the ⊙, the value of the ∠ X is supplementary to that found above, and the conjugate segment is obtained.

43. Let PA, QB intersect at X. Join PB. [In the fig. taken PQ and AB do not intersect within the circle, and X is also external].

Then ∠ X is the diff. of ∠ˢ PBQ, XPB [I. 32].

But ∠ PBQ is constant, being a rt. angle [III. 31].

Also ∠ XPB is constant, being subtended by the fixed arc AB.

∴ the ∠ X is constant; and since the points A, B are fixed, the locus of X is the arc of a segment [III. 21]. When X is internal, the ∠ X is supplementary to the value found above, and the conjugate segment of the locus is obtained.

44. It follows that AP = AC, ∴ ∠ APC = ∠ ACP.

But ∠ BAC = sum of ∠s APC, ACP [I. 32].

∴ ∠ BAC is double ∠ APC.

Or, ∠ BPC is half of ∠ BAC, and is therefore constant.

Then, since BC is fixed, the locus of P is the arc of a segment on BC [III. 21, Cor.].

45. The intersection of the diagonals is X, the middle point of BC [Ex. 5, p. 70]. Join X to D, the middle point of AB.

Then XD is par^l. to AC [Ex. 2, p. 104].

∴ ∠ DXB = ∠ ACB [I. 29].

But ∠ ACB is constant [III. 21].

∴ ∠ DXB is constant, and D, B are fixed points.

∴ the locus of X is a ⊙, the segment on DB being similar to the segment ACB.

46. Let A be the point of intersection of the rulers.

Then PXQA is a rectangle.

∴ AX = PQ, which is constant, and the point A is fixed. Hence the locus of X is the quadrant of a circle described from the centre A with radius PQ.

47. Prove as in [Ex. 9, p. 234] that ∠ PXQ = ∠ CAD, (or is supplementary to it).

But C, A, D are fixed points; and the arms PX, QX pass through two fixed points C, D.

∴ the locus is a ⊙ through C and D.

And since ∠ CBD = ∠ CAD [I. 8],

∴ this ⊙ passes through B.

48. [Take the figure in which PA and PB must both be produced to meet the second ○^{ce}.] Let AY, BX intersect at R.

Then the locus of R is required.

Now ∠ ARB = sum of ∠s RBY, RYB [I. 32]
= sum of ∠s at P, X, Y [I. 32],
and these are all constant, being subtended by fixed arcs.

∴ ∠ ARB is constant; and since the points A and B are fixed, the locus is part of a circle. If PA or PB cuts the ○ce without being produced, the ∠ ARB = the supplement of the sum of the ∠s P, X, Y. Hence the rest of the circle is obtained.

49. Let PH and KQ intersect at X. Required the locus of X.

From the △ PXQ it will be seen by I. 32 that the ∠ X = the diff. of the ∠s HPA, AQK; both of which are constant, since they stand on the fixed arcs HA, AK.

And since H, K are fixed points, the locus of X is part of a ⊙.

If P and Q are on the same side of A, the value of the ∠ X is supplementary to that found above, and the rest of the ⊙ is obtained.

50. Let the bisectors meet at X. Then the locus of X is required.

Now ∠ XAB = one-half of sum of ∠s PAB, QAB.

And ∠ XBA = one-half of sum of ∠s PBA, QBA.

∴ the sum of the ∠s at the base of △ XAB = one-half of the sum of the ∠s at the base of △s PAB, QAB.

Hence [I. 32] the vertical ∠ AXB = one-half of vertical ∠s APB, AQB, both of which are constant [III. 21].

∴ ∠ AXB is constant; and A, B are fixed points. ∴ the locus of X is the arc of a segment of ⊙ on base AB [III. 21, Cor.].

51. Let C, D be the centres of the two ⊙s, and in the figure considered let X, the middle point of PQ, fall in PA.

Bisect CD at G, and draw CE, GH, DF perp. to PQ.

Then EF = $\frac{1}{2}$PQ; for EA = $\frac{1}{2}$PA, and AF = $\frac{1}{2}$AQ.

∴ EF = XQ: also EH = HF [Ex. 14, p. 106].

Hence it may be shewn that XH = HA.

Then from the △s GHX, GHA, we have GX = GA [I. 4].

∴ the locus is a circle, with centre G and radius GA or GB.

A better proof follows from Book VI., Prop. 6.

Join BP, BX, BQ. Then for all positions of PQ the angles of the △ BPQ are constant [III. 21 and I. 32].

∴ the ratio BP : PQ is constant [VI. 4]: hence the ratio BP : PX is constant.

But the ∠ BPX is constant: hence [VI. 6] the ∠ PXB is constant.

∴ the ∠ BXA is constant. ∴ the locus of X is the arc of a segment on AB.

Miscellaneous Examples on Angles in a Circle.

Page 248.

52. Because the points P, Q, C, B are concyclic;
∴ the ∠ˢ BPQ, BCQ together = two rt. angles [III. 22].
Similarly, the ∠ˢ BP′Q′, BCQ′ together = two rt. angles.
∴ ∠ BPQ = ∠ BP′Q′; ∴ PQ and P′Q′ are par[l]. [I. 28].
Again, let TAT′ be the tangent at A to the circum-⊙.
Then ∠ TAB = ∠ BCA [III. 32].
Hence ∠ BAT′ = ∠ BPQ [I. 13 and III. 22].
∴ TT′ is par[l]. to PQ.

53. [In the fig. taken AB, AC *when produced* meet the second ⊙ at D and E].

Let AT be the tangent at A: then
∠ TAB = ∠ ACB [III. 32]
= ∠ BDE [Ex. 5, p. 241];
∴ TA is par[l]. to DE [I. 27].

54. For ∠ PTA = ∠ TBA [III. 32]; and ∠ ATC = ∠ CTB;
hence ∠ PTC = sum of ∠ˢ PTA, ATC
= sum of ∠ˢ TBC, CTB
= ext. ∠ TCP [I. 32];
∴ PT = PC [I. 6].

55. Join BF. Then ∠ BCA = ∠ BFA [III. 21]
= comp[t]. of ∠ BAF [III. 31]
= ∠ ADF.

∴ ∠ BCA = ∠ ADE, and the ∠ DAE is common to the two △ˢ; ∴ ∠ ABC = ∠ AED [I. 32].

56. Join AD. Then the points B, F, O, D are concyclic;

∴ ∠ BOD = ∠ BFD [III. 21]

= sum of ∠s FAD, FDA [I. 32].

Similarly ∠ COD = sum of ∠s EAD, EDA.

Hence, by addition, ∠ BOC = sum of ∠s BAC, FDE.

57. Let A be the external point, BC the chord of contact, and let the tangent AB be produced to D.

Then ∠ BAC = the diff. of ∠s DBC, BCA [I. 32]

= the diff. of ∠s in the alt. segments. [III. 32].

58. Let A be the point of intersection of the two ⊙s, AD, AE the two diams., and let the line through A meet the ⊙s at X and Y.

Then in the △s AXD, AYE

∠ DAX = ∠ EAY [Hyp.], and ∠ AXD = ∠ AYE [III. 31];

∴ ∠ ADX = ∠ AEY [I. 32].

∴ the segments are similar.

59. Let ABX, ABY be the two equal ⊙s, and let the ⊙ described from centre A cut the ⊙ ABY at C and the ⊙ ABX at D, the points C, D being on the same side of AB.

Then the arc AC = the arc AD, for they are cut off from equal ⊙s by equal chords; and B is a point on the ○ce of both of the given ⊙s; hence the arcs DA, AC subtend equal angles at B on the same side of AB [III. 27]. That is, BC and BD coincide in direction; or, the points B, C, D are collinear.

60. [In the fig. taken the △ ABC is acute angled, and A′ is on the minor arc AB].

Because B′A′, A′C′ are parl. respectively to BA, AC,

∴ the ∠ A′ = the ∠ A, ∴ the arc B′C′ = the arc BC [III. 26]

From these equal arcs take the arc BC′:

then the arc BB′ = the arc CC′;

∴ the ∠ B′CB = the ∠ CBC′ [III. 27];

∴ B′C is parl. to BC′ [I. 27].

61. Join HB, BK, AB.

Then $\angle$ HBK + $\angle$ X = $\angle$ ABK + $\angle$ ABH + $\angle$ X
= $\angle$ ABK + $\angle$ QPX + $\angle$ X [III. 21]
= $\angle$ ABK + $\angle$ AQK [I. 32]
= two rt. angles [III. 22];

∴ the points H, B, K, X are concyclic [III. 22. Converse].

62. Let AB be the given st. line, P the given point of contact, and X and Y the given points in AB.

[The problem is only possible when P is between X and Y.]

At P draw PQ perp. to AB; then the centre of the required ⊙ lies on PQ.

On XY describe a semicircle, meeting PQ at O.

From centre O, with radius OP, describe a ⊙, and from X and Y draw the tangents XC, YD. These tangents shall be par[l].

This is proved by shewing by the converse of [Ex. 10, p. 197] that the sum of the $\angle$s CXP, DYP is two rt. angles.

63. Because the $\angle$s CXP, CYP are rt. angles,

∴ the four points C, X, P, Y lie on a ⊙ whose diameter is CP [III. 31]. And this ⊙ is of constant magnitude, since CP is a radius of the given ⊙.

Now the $\angle$ YCX is also constant. ∴ the chord XY is constant.

64. Call the tangent NPT. Join AP.

Then the $\angle$ BPT = the $\angle$ PAB [III. 32]
= the $\angle$ MNP [III. 21],

for the points A, N, P, M are obviously concyclic.

Hence MN and PB are par[l]. [I. 28].

65. Join XN, YN.

Then each of the $\angle$s AXN, APB, NYB is a rt. angle [III. 31].

∴ the fig. XNPY is a rectangle.

∴ $\angle$ NXY = $\angle$ NPY
= $\angle$ NAX, from the rt. angled $\triangle$s PAB, NPB [I. 32].

∴ XY touches ⊙ AXN [Converse of III. 32].

Or, *otherwise*. Join X to C, the centre of the ⊙ AXN.

Then $\angle$ CXA = $\angle$ CAX = $\angle$ NPB = NXY.

Hence ∠ AXN = ∠ CXY.

∴ ∠ CXY is a rt. angle; ∴ XY is a tangent.

Similarly XY may be proved a tangent to the other circle.

66. Let AB be the common chord, through A draw APXQ to cut the arcs. Then shall PX = QX.

For since ∠ APB is the supp^t. of ∠ AQB,

∴ ∠ BPQ = ∠ AQB,

and ∠^s BXP, BXQ are rt. angles [III. 31].

Hence PX = QX [I. 26].

67. Let AD, AE be the given lines touching the given ⊙ at B and C. Let the chord PQ be bisected by BC at Z, and produced to meet AD and AE at X and Y.

Then shall PX = QY.

Take centre O. Join OZ, OB, OC, OX, OY.

Then the ∠^s OZX, OBX are rt. ∠^s [III. 3, III. 18];

∴ the four points O, Z, B, X are concyclic [III. 22].

∴ the ∠ ZXO = the ∠ ZBO, in the same segment.

Similarly, the ∠ ZYO = the ∠ ZCO.

But since OB = OC, ∴ the ∠ ZBO = the ∠ ZCO.

∴ ∠ ZXO = ∠ ZYO, ∴ ZX = ZY [I. 6].

And by hyp. ZP = ZY, ∴ PX = QY.

68. Let C, D be the centres of the given ⊙^s which intersect at A, and X the given line.

On CD describe a semicircle; and from centre D with radius half of X cut this semicircle at E. Join ED.

Through A draw PAQ par^l. to ED. PQ shall be the line required. Join CE and produce it to meet PQ at G, and draw DH par^l. to CG, meeting PQ at H.

Then since ∠ CED is a right ∠ [III. 31],

∴ CG, DH are perp. to PQ [I. 29], and GH = ED.

Also GH is half of PQ [III. 3];

∴ PQ = X, and is drawn through A.

69. Let ABC be the given △, on the sides of which equilat. △^s are described externally, and let the ⊙^s about the equilat. △^s on BC, CA meet at O. Join AO, BO, CO.

[In the fig. taken O falls within the △.]

Since the ∠ of an equilat. △ is $\frac{1}{3}$ of two rt. angles,

∴ each of the ∠ˢ AOC, BOC is $\frac{2}{3}$ of two rt. angles [III. 22].

Hence the ∠ AOB is $\frac{2}{3}$ of two rt. angles [I. 15, Cor. 1].

∴ a circle described about the equilat. △ on AB will pass through O [III. 22. Converse].

70. Let the ⊙ˢ about the △ˢ BRP, CPQ intersect at O.

Join PO, RO, QO.

Then ∠ POR = supplement of ∠ B
Also ∠ POQ = supplement of ∠ C } [III. 22].

But since the three ∠ˢ POR, POQ, ROQ = 4 rt. angles, and A + B + C = 2 rt. angles,

∴ ∠ ROQ = supplement of ∠ A.

∴ a ⊙ about △ RAQ will pass through the point O
[III. 22. Converse].

71. On each of the sides of the △ describe segments containing an angle equal to $\frac{2}{3}$ of two right angles (twice the ∠ of an equilat. △).

Then [Ex. 69] the arcs of these segments meet at a point, at which each side will subtend an angle equal to $\frac{2}{3}$ of two right angles, or $\frac{1}{3}$ of *four* rt. angles.

72. Let P, Q, R be the fixed points. On PR and PQ describe (externally to the △ PQR) segments containing an angle of an equilat. △.

Through P draw *any* st. line BC terminated by the ○ᶜᵉˢ.

Join BR, CQ, and produce them to meet at A.

Then ABC is an equilat. △.

For since each of the ∠ˢ B, C is one-third of two rt. angles,

∴ the ∠ A is also one-third of two rt. angles.

73. Let P, Q, R be the given points, ABC the given △. On PQ, RP describe segments (externally to the △ PQR) capable of containing angles equal to the ∠ˢ B, C. Through P draw B′PC′ terminated by the ○ᶜᵉˢ and equal to BC [Ex. 68, p. 249]. Join B′Q, C′R, and produce them to meet at A′. Then A′B′C′ shall be the required △. [III. 21 and I. 26.]

SIMSON'S LINE. **Page 250.**

75. Take the figure of Ex. 74, p. 250.

Now as in Ex. 74, the $\angle$ PDF = the $\angle$ PBF = $\angle$ ACP.

Hence the $\angle^s$ PDF, PDE = the $\angle^s$ ECP, PDE.

But the $\angle^s$ PDF, PDE = two rt. angles [I. 13].

∴ the $\angle^s$ ECP, PDE = two rt. angles.

∴ the points P, D, E, C are concyclic [III. 22. Converse].

∴ the $\angle$ PEC = the $\angle$ PDC, in the same segment,

= a rt. angle [Constr.].

76. Again take the figure of Ex. 74, p. 250.

Since the points P, F, B, D are concyclic,

∴ the $\angle$ DPB = the $\angle$ DFB, in the same segment

And since the points P, C, E, D are concyclic,

∴ the $\angle$ DPC = the supplement of $\angle$ DEC [III. 22]

= the $\angle$ DEA.

By addition, the whole $\angle$ BPC = the sum of the $\angle^s$ EFA, FEA

= the supplement of the $\angle$ A.

∴ P lies on the $\bigcirc^{ce}$ of the ⊙ circumscribed about the $\triangle$ABC.

77. Draw PD, PD′, PE, PF perp. respectively to the four lines BC, B′C′, ACC′, ABB′.

Then since P is on the ⊙ circumscribed about the $\triangle$ABC, ∴ the points E, F, D are collinear [Ex. 74].

And since P is on the ⊙ circumscribed about the $\triangle$AB′C′,

∴ the points E, F, D′ are collinear.

Hence D and D′ both lie on the st. line through E and F.

78. Let ABC be the $\triangle$, P the given point on the circumscribed ⊙.

Let PF, PD be the perps. on AB, BC; so that FD produced is the pedal of P. Draw AH perp. to BC, and produce it to meet the $\bigcirc^{ce}$ at G. Take HO equal to HG. Then O is the orthocentre [Ex. 21, p. 244].

Let OP meet the pedal of P at X. Then shall OX = XP.

Draw PB, PC. Let PG, produced if necessary, meet the pedal at K, and BC (or BC produced) at L. Join OL.

[The proof given below is for an acute-angled triangle. P is taken in the arc BG. F falls within AB, and PG meets BC produced.]

Then ∠ PDK = ∠ PBF [Ex. 5, p. 241] = supp^t. of ∠ ACP [III. 22]
= supp^t. of ∠ AGP [III. 21] = ∠ DPG [I. 29].

So that ∠ KDL = ∠ KLD, since △ PDL is rt. angled [I. 32].

∴ PK = KD = KL.

But by I. 4, the △^{s} HLG, HLO are equal in all respects.

∴ ∠ OLH = ∠ DLK = ∠ KDL; ∴ XK, OL are par^l. But K is the middle point of PL, ∴ X is the middle point of OP [Ex. 1, p. 104].

IV. On the Circle in connection with Rectangles.

Page 251.

2. Since the ∠^{s} AEB, ADB are rt. angles, ∴ the four points A, E, D, B are concyclic [III. 31],

∴ the rect. AO, OD = the rect. BO, OE [III. 35].

Similarly, the rect. BO, OE = the rect. CO, OF.

3. Since the ∠^{s} AEB, ADB are rt. angles,

∴ the points A, E, D, B are concyclic;

∴ the rect. CA, CE = the rect. CB, CD [III. 36 Cor.].

4. Since the ∠^{s} ADE, ACE are rt. angles,

∴ the four points D, A, C, E are concyclic [III. 22]

∴ BE . BC = BD . BA [III. 36],

or, BE . EC + BE^2 = BD . DA + BD^2 [II. 3],

or, BE^2 – BD^2 = BD . DA – BE . EC,

that is, DE^2 = BD . DA – BE . EC [I. 47].

5. Let any ⊙ through P, Q cut the given ⊙ at R.

Then by Ex. 1, p. 251, the rect. $OP . OQ = OR^2$.

∴ OR is a tangent to the second ⊙ [III. 37].

Hence the tangents to the two ⊙s at the point R are perp. to one another; ∴ the circles cut orthogonally.

6. Let A, B be the given points through which all the circles pass, and C the fixed point in BA produced.

From C draw CT a tangent to *any one* of the circles.

Then $CT^2 = CA . CB$ [III. 36].

∴ CT is constant. That is, all the points of contact are at a constant distance from the fixed point C; ∴ their locus is a ⊙ with centre C. And since each radius of this ⊙ is a tangent to a ⊙ of the given series, ∴ the locus cuts each ⊙ of the system orthogonally.

7. Let C be the centre of the given ⊙, and A the given fixed point. Let DAT be *any* ⊙ passing through A, and cutting the given ⊙ orthogonally at T. Join CA, and produce it, if necessary, to meet the ⊙ DAT at B.

Then since the ⊙s cut orthogonally at T, CT is a tangent to the ⊙ DAT [III. 16].

$$\therefore CB . CA = CT^2 \text{ [III. 36]}.$$

But CA and CT are constant; ∴ CB is constant.

∴ B is a fixed point.

8. Since by the last Ex. all ⊙s which pass through a fixed point A and cut a given ⊙ orthogonally pass also through a second point B, the locus of their centres is the st. line bisecting AB at rt. angles [III. 1].

To find this point B. Draw *any* radius CT to the given ⊙.

Describe a ⊙ to pass through A and touch CT at T [Ex. 28, p. 238].

This ⊙ will cut the given ⊙ orthogonally. Join CA, and produce it if necessary, to cut the ⊙ of construction at B. Then B is the required point.

9. Let C be the centre of the given ⊙, and A, D the given points.

Then by Ex. 7, all ⊙s through A cutting the given ⊙ orthogonally must pass through another fixed point B. Find B, as in the last Example. Then the ⊙ circumscribed about ABD is that required.

10. Describe any ⊙ to pass through the points A, B, and any other ⊙ through C, D intersecting the first at X, Y.

Join XY, and produce it to meet AD at O. Then O is the required point.

For OA . OB = OX . OY [III. 36]
= OC . OD [III. 36].

11. Let AB and CD intersect at E. Join BQ.

Then the ∠s PQB, PEB are rt. angles [III. 31, and Hyp.]

∴ the points Q, P, B, E are concyclic;

∴ AQ . AP = AE . AB [III. 36].

And since AE and AB are constant,

∴ rect. AQ, AP is constant.

12. Let CD cut AB at E. Join BQ, BC.

Then the ∠s PEB, PQB are rt. angles [Hyp., and III. 31].

∴ the points E, P, Q, B are concyclic;

∴ AP . AQ = AE . AB [III. 36].

Now the ⊙ about the △CEB has its centre on BC,

for the ∠CEB is a rt. angle [III. 31].

And AC is perp. to BC [III. 31], ∴ AC is a tangent to the ⊙ about the △CEB.

Hence AE, AB = AC^2 [III. 36],

∴ AP . AQ = AC^2.

13. Draw AE perp. to CD, and from Q draw QR perp. to AP meeting AE at R.

Then since the ∠s PQR, PER are rt. angles,

the points P, E, R, Q are concyclic

∴ AE . AR = AP . AQ [III. 36].

∴ AE . AR is constant, for AP . AQ is constant [Hyp.]; and since AE is constant, ∴ AR is constant. That is, R is a fixed point.

And the ∠ AQR is a rt. angle.

∴ the locus of Q is a circle on AR as diameter.

14. Let T be one point of intersection of the two given ⊙s, A any point on the ○ce of one of them, and C the centre of the other. Draw AC, and produce it if necessary to meet the ○ce of the first ⊙ at B. Join CT.

Then since the ⊙s are orthogonal, CT is a tangent;

$$\therefore CT^2 = CA . CB \quad [\text{III. } 36].$$

Hence [Ex. 1, p. 251] B is the point at which AC is cut by the chord of contact of tangents from A.

But this chord is bisected at rt. angles by AC.

Hence the first ⊙ passes through its middle point.

15. Draw PX perp. to AB. Join AD, BC.

Then since the ∠s PCB, PXB are rt. angles [III. 31, and Constr.],

∴ the points P, C, B, X are concyclic;

$$\therefore AP . AC = AX . AB \quad [\text{III. } 36].$$

Similarly $BP . BD = BX . BA$;

$$\therefore AP . AC + BP . BD = AX . AB + BX . AB$$
$$= AB^2 \quad [\text{II. } 2].$$

16. For by Ex. 3, p. 229, GA = GE, and HD = HF.

Also by Ex. 19, p. 237, AE = DF. Hence it may be proved that

$$GB = HC.$$

Now since GC is divided at B,

$$\therefore 4GC . GB + BC^2 = GH^2 \quad [\text{II. } 8],$$

or,

$$4GA^2 + BC^2 = GH^2 \quad [\text{III. } 36];$$

that is,

$$AE^2 + BC^2 = GH^2.$$

17. Let PM meet the ○ce of the given ⊙ at XY.

Then because XY is bisected at M and produced to P,

$$\therefore PM^2 = PX . PY + XM^2 \quad [\text{II. } 6]$$
$$= PC . PD + AM . MB \quad [\text{III. } 36, \text{ and } 35].$$

18. Join AF, AG. Then shall AF, AG be in the same st. line. Join DB, DC.

[Various figures arise according to the magnitude and disposition of the given ⊙^{s}; the proof given below may be adapted to the various cases by interchanging the application of III. 21 and III. 22.]

(i) Because the four points G, A, C, D are concyclic,

∴ the ∠ GAD = ∠ GCD [III. 21].

And because the points F, A, B, D are concyclic,

∴ the ∠ FAD = ∠ FBD [III. 21].

∴ the ∠^{s} GAD, FAD = the ∠^{s} ECD, EBD

= two rt. angles, for the points E, B, D, C are concyclic [III. 22].

∴ GA, AF are in the same st. line.

(ii) EF . EB = EA . ED = EC . EG [III. 35 or 36];

∴ the points B, F, C, G are concyclic.

19. Join AO, and produce it to meet BC at D.

Then $AB^2 + AC^2 = AB \cdot AF + AB \cdot BF + AC \cdot AE + AC \cdot CE$ [II. 2].

But AB . AF = AC . AE = the sq. on tangent from A [III. 36].

And it may be proved that $AB \cdot AF + AC \cdot AE = BC^2$ [Ex. 15, p. 252].

∴ $AB^2 + AC^2 = BC^2$ + twice the sq. on the tangent from A.

20. Let AB be the given diameter.

Then $PQ^2 = PX^2 + QX^2 + 2PX \cdot XQ$ [II. 4],

∴ $PQ^2 + PY^2 + QY^2 = PX^2 + PY^2 + QX^2 + QY^2 + 2PX \cdot XQ$.

But $PX^2 + PY^2 = 2PC^2 + 2XC^2$,
and $QX^2 + QY^2 = 2QC^2 + 2XC^2$ } which are constant.

Also PX . XQ = AX . XB [III. 35], which is constant;

∴ $PQ^2 + PY^2 + QY^2$ is constant.

Problems on Tangency. Page 257.

26. Let A be the given point, BC the st. line to be touched, and PQ the line on which the centre is to lie.

From A draw AM perp. to PQ, and produce AM to A′, making MA′ equal to MA.

Then the required ⊙ must pass through A′ [Ex. 1, p. **233**].

Hence we have only to describe a ⊙ through A, A′ to touch BC. This is done in Ex. 21, p. 253.

27. As in the last example a second point may be found through which the required ⊙ must pass [Ex. 1, p. **233**].

The problem is thus reduced to that solved in Ex. 22, p. 254.

28. Let A and B be the given points, and C the centre of the given ⊙, of which XY is a given arc. Required to describe a ⊙ to pass through A, B and cut off from the given ⊙ an arc equal to XY.

Join XY, and from centre C describe a circle to touch XY. Then the given ⊙ intercepts on every tangent to the ⊙ of construction a part equal to XY.

Describe a ⊙ through A, B to touch the given ⊙ at T

[Ex. 21, p. **253**].

Draw the common tangent at T to meet BA produced at O. From O draw a tangent to the ⊙ of construction cutting the given ⊙ at P, Q.

Then $OA . OB = OT^2 = OP . OQ$ [III. 36].

∴ A, B, P, Q are concyclic.

∴ a ⊙ described through the points A, B, P will pass through Q.

And PQ = XY [Ex. 5, p. 195], ∴ arc PQ = arc XY [III. 28].

29. Worked out on page 239 of Euclid.

30. Let A, B be the centres of the given ⊙^{s}, and PQ the given st. line. Required to draw a ⊙ to touch the given ⊙^{s} and PQ. Of the two ⊙^{s} (A), (B) let (B) be the greater. From centre B, with radius equal to the difference of the given ⊙^{s}, describe a ⊙.

Draw XY par[l]. to PQ, at a distance from it equal to the radius of (A), and on the side remote from the given ⊙^{s}.

Through A describe a ⊙ (centre O) to touch the ⊙ of construction at D and XY at F [Ex. 25, p. 256].

Join OA, OB, OF meeting the given ⊙^{s} and PQ respectively at G, E, C. Then clearly

$$OG = OE = OC.$$

∴ a ⊙ described from centre O with radius OC is that required.

V. On Maxima and Minima.

Page 262.

1. Take one of the given sides as base, and from an extremity of this base, with the other side as radius, describe a ⊙.

Then the vertex of the △ must lie on this ⊙. Now, the base being given, the △ is greatest when the altitude is greatest: this may readily be shewn to be when the second side is drawn at rt. angles to the first [I. 19].

2. If the base and area of a △ABC are given, the vertex C must move on a st. line PQ par[l]. to the base AB [I. 39].

And since AB is fixed, the perimeter is least when the sum of AC, BC is least. This is the case when

the ∠PCA = the ∠QCB [Ex. 3, p. 261].

But ∠PCA = ∠CAB, and ∠QCB = ∠CBA [I. 29].

∴ ∠CAB = ∠CBA; ∴ AC = BC [I. 6].

3. If the base and vertical ∠ of a △ are given, then the vertex must move on the segment of a ⊙ described on the base BC, and containing the given angle [III. 21. Cor.]; and the △ of greatest area is that which has the greatest altitude.

Now the greatest altitude is the st. line AX which bisects BC at X at rt. angles. For take *any* other point P on the arc of the segment, and draw PM perp. to BC. Find O, the centre, which must lie in AX or AX produced [III. 1]. Join OP, and let PM (produced, if necessary) meet the diameter par[1] to BC at N.

Then OP is greater than PN [I. 19]; ∴ XA is greater than PM, since OX = NM.

4. Let O be the centre of the given ⊙, and MN the given st. line.

Draw OA perp. to MN, and let P be any point in MN. Then the tangents from A shall contain a greater angle than the tangents from P.

From A and P draw tangents AB, PQ. Join OB, OQ, OP.

Then OP is greater than OA [I. 18].

And $OP^2 = OQ^2 + PQ^2$; also $OA^2 = OB^2 + AB^2$.

But OB = OQ. Hence PQ is greater than AB.

In QP make QX equal to BA, and join OX.

Then the △s ABO, XQO are identically equal [I. 4].

∴ ∠ BAO = ∠ QXO. But ∠ QXO is greater than ∠ QPO [I. 16].

∴ ∠ BAO is greater than ∠ QPO.

But the ∠ between the tangents at A is double the ∠ BAO, and the ∠ between the tangents at P is double the ∠ QPO.

∴ the tangents from A include the greater angle.

5. Let AB be the straight rod, C its middle point, and O the intersection of the rulers. Then OC = half of AB and is constant for all positions of AB [III. 31].

And since the base is given in magnitude, the area of the △ is greatest when the perp. from O on AB is the greatest.

Now when AB makes equal angles with the two rulers, OC is perp. to AB [I. 6, and 4]. And in any other position of AB, OC is greater than the perp. from O on AB [I. 19].

Hence the greatest △ is obtained when AB is equally inclined to the two rulers.

6. Let AB be the given line, and K the side of the given square.

(i) At B draw BC, making the ∠ ABC half a rt. angle.

From centre A, with radius K, describe a ⊙ cutting BC at P (or P′). Draw PX perp. to AB.

Then shall $AX^2 + XB^2 =$ sq. on K.

For ∠ XBP $= \frac{1}{2}$ rt. angle, and ∠ PXB = one rt. angle,

∴ ∠ XPB $= \frac{1}{2}$ rt. angle; ∴ PX = XB.

Hence $AX^2 + XB^2 = AX^2 + XP^2$

$= AP^2 =$ sq. on K [I. 47].

(ii) Thus $AX^2 + XB^2$ is a minimum, when AP is a minimum; that is, when AP is the perp. on BC.

In this case the ∠ PAB $= \frac{1}{2}$ rt. angle = ∠ ABP.

∴ AP = BP, and hence X is the middle point of AB.

7. (i) See Ex. 68, p. 249.

(ii) Using the letters there employed, we see that PAQ is a maximum, when ED is a maximum. But ED has its greatest value when it coincides with CD. Hence PQ is a maximum when it is par[l]. to CD.

8. Let OA, OB be the tangents. Take P the middle point of the arc AB, and let PX, PY be the perps. on OA, OB.

Then PX + PY shall be a minimum.

Let Q be any other point on the arc AB (in the fig. chosen, Q is on PB), and QM, QN the perps. on OA, OB. Let PY, QM intersect at R. Join PQ.

Then the tangent at P may be shewn to make equal ∠s with OA, OB, therefore with PY, QM. Hence if this tangent cuts QM at K, K must be without the ⊙, and the ∠ RKP = the ∠ RPK.

But ∠ RKP is greater than ∠ RQP [I. 16],

∴ ∠ RPQ is greater than ∠ RQP.

∴ RQ is greater than RP.

Also RM = PX, and QN = RY.

∴ RQ + RM + QN is greater than RP + PX + RY,

or QM + QN is greater than PX + PY.

9. Let A and B be the fixed points, PQ the tangent at T, and let ∠ PTA = ∠ QTB.

Then AT + BT shall be a minimum.

This problem supposes that AB does not meet the ⊙, and that AT, BT are on the side of PQ remote from the ⊙.

Take X any other point on the ○ce: then AX must cut PQ (hyp.) at some point K. Join KB, XB.

Then AK + KB is greater than AT + TB [Ex. 3, p. 261];

and AX + XB is greater than AK + KB [I. 21].

Hence AX + XB is greater than AT + TB.

10. Let AP, AQ be st. lines of indefinite length including the fixed vertical angle.

Let ABC be the isosceles △, having the given altitude AD.

And let AB′C′ be any other △ having an equal altitude AD′.

Then by I. 26, D is the middle point of BC.

Through D draw XDY par^l. to B′C′ meeting AP, AQ at X, Y.

Now △ABC is less than △AXY [Ex. 4, p. 262].

And △AXY is less than △AB′C′ by the strip B′XYC′, since it may be shewn that B′C′ must lie on the side of XY remote from A. For let XY meet AD′, or AD′ produced, at K: then the ∠ AKD is a rt. angle, ∴ AD is greater than AK; that is, AD′ is greater than AK.

11. For all such triangles have the same vertical angle, and the same altitude, namely the radius of the ⊙ [III. 18].

And it may easily be shewn that the △ whose base is bisected at the point of contact is isosceles [I. 4]; hence the proof of the last exercise applies.

12. Let AP and AQ be the two fixed tangents, and BC any other tangent to the convex arc. Let O be the centre of the ⊙. Join OB, OC.

Then the quad^l. APOQ is of constant area.

Hence the △ABC is a maximum when the fig. OPBCQ is a minimum.

And this figure is double the △OBC [proved as in Ex. 6, p. 235].

Hence △ABC is a maximum, when △BOC is a minimum.

Now the ∠BOC is constant [Ex. 6, p. 235], and the altitude of the △OBC is also constant, ∴ its area is a minimum when it is isosceles [Ex. 10, p. 263]; that is, when BC touches the arc PQ at its middle point.

13. Let AB be the given base. Then, since the area is given, the vertex must lie on some st. line XY par[l]. to AB. Describe a ⊙ to pass through A, B and to touch XY at C [Ex. 21, p. 253, note]. Let P be any other point in XY.

Join AC, CB and AP, PB.

Then one at least of the lines AP, BP must cut the ⊙. Let AP cut it at Q. Join BQ.

Then ∠AQB is greater than ∠APB [I. 16]

and ∠AQB = ∠ACB [III. 21];

∴ ∠ACB is greater than ∠APB.

And from the construction of the ⊙ it may be proved that AC = BC [I. 4].

14. Let A, B be the given points (both without the given ⊙). Through A and B describe a ⊙ to touch the given ⊙ externally at C [Ex. 22, p. 254].

Then ACB shall be the maximum angle. Let P be any other point on the ○[ce] of the given ⊙. Join AP, BP, and let AP meet the ⊙ of construction at Q. Join QB.

Then ∠AQB is greater than ∠APB [I. 16]

and ∠AQB = ∠ACB [III. 21].

∴ ∠ACB is greater than ∠APB.

If two circles can be drawn so as to be touched externally by the given circle, two points of maximum angle can be found, one on each side of AB.

15. Let ABCD be the bridge, where AB = 49 ft., BC = 32 ft., and CD = 49 ft. The st. line AP represents the bank.

Through B, C describe a ⊙ to touch AP at T [Ex. 21, p. 253].

Then the arch BC subtends the greatest angle at T [Ex. 2, p. 260].

Also $$AT^2 = AB \, . \, AC \quad [\text{III. } 36]$$
$$= 49 \, . \, 81 \text{ sq. ft.}$$
$$\therefore \; AT = 7 \times 9 = 63 \text{ ft.}$$

16. Since the sides AC, BC are constant, the area of the △ABC is a maximum when they are at rt. angles [Ex. 1, p. 262].

Draw any two radii CA′, CB′ at right angles, and join A′B′.

Then, by Ex. 11, p. 235, through P draw the st. line PAB so that the part AB intercepted by the ○ce may be equal to A′B′. Then clearly the ∠ ACB is a rt. angle [I. 8].

17. Let ABCD be a rectangle inscribed in a given ⊙.

Join AC. Then AC is a diameter [III. 31].

Now the rectangle is double of the △ABC.

And since the base AC is constant, the △ABC is greatest when the altitude BX, namely the perp. from B on AC, is greatest.

And BX may be shewn to be greatest when B is the middle point of the arc AC. The rectangle then becomes a square.

18. Let O be the centre of the given ⊙. Bisect AB at X, and join XO cutting the ○ce at P. Join AP, PB.

Then shall $AP^2 + PB^2$ be a minimum.

Now $AP^2 + PB^2 = 2AX^2 + 2XP^2$ [Ex. 24, p. 161].

Hence, since AX is constant, $AP^2 + PB^2$ is a minimum when XP is a minimum.

But XP is the least of all st. lines drawn from X to the ○ce [III. 8].

19. It is shewn in Ex. 4, p. 223 how to find a point C such that AC + PC may be equal to a given line H. Now the greatest value H can have, in order that this construction should be possible, is the diameter of the second segment. This determines the point X, and therefore the point C: and it may easily be shewn by III. 31 that CX = CB = CA; that is, that C is the middle point of the arc AB.

20. No inscribed triangle that is not equilateral can have the maximum perimeter.

For let PQR be an inscribed triangle not equilateral; then it must have one pair of sides unequal, say PQ, QR. Hence there is an inscribed △ on the base PR, which has a greater perimeter [Ex. 19, p. 264], ∴ the △PQR is not the inscribed △ of greatest perimeter. And this argument may be applied to *all* inscribed triangles not equilateral.

21. No inscribed triangle that is not equilateral can have the greatest area.

For let PQR be an inscribed △ not equilateral; then it must have one pair of sides unequal, say PQ, QR. Hence [Ex. 3, p. 262] there is an inscribed △ on the base PR, which has a greater area.

∴ the △PQR is not the inscribed △ of greatest area.

And this argument may be applied to *all* inscribed △s not equilateral.

22. It has been proved [Ex. 20, p. 243] that every two sides of the pedal triangle are equally inclined to that side of the original triangle, in which they meet.

Also [Ex. 3, p. 261] if A and B are fixed points and P a point in a given st. line CD, then AP + PB is a minimum, when these lines are equally inclined to CD.

Thus no triangle inscribed in the △ABC, that is not the pedal triangle, can have the minimum perimeter.

For let PQR be an inscribed △, not the pedal △. Then at least one pair of its sides, say PR, QR, are not equally inclined to the side AB in which they meet. Hence there is an inscribed △ on the base PQ which has a less perimeter [Ex. 3, p. 243].

∴ the △PQR is *not* the inscribed △ of least perimeter. And this argument may be applied to all inscribed △s, except the pedal △.

23. Adopting the figure of II. 14.

The sq. on EH = the rectangle BD in area.

Now since EF = ED, it follows that BF is half the perimeter of the rectangle ; ∴ GH is one-quarter of the perimeter of the rectangle. Also HE is one-quarter of the perimeter of the square.

But GH is greater than HE [I. 19].

Hence the perimeter of the square is less than the perimeter of the rectangle.

24. Let D, E, F be the fixed points, and XYZ the given △. Join FD, DE, and on these lines describe segments containing the ∠ˢ Y, Z respectively.

Through D draw the maximum line BC terminated by the two ○ᶜᵉˢ [Ex. 7, p. 263]. Join BF, CE; and produce them to meet at A.

Then since the ∠ˢ B, C are respectively equal to the ∠ˢ Y, Z, ∴ the remaining ∠ A = remaining ∠ X [I. 32].

And since the ∠ˢ of the △ABC are fixed, the area is a maximum, when any one of its sides is a maximum. But BC is a maximum [Constr.].

∴ the △ ABC is a maximum.

VI. Harder Miscellaneous Examples.

Page 264.

1. For let O be the centre of the given ⊙. Join OC.

Then ∠ EDC = ∠ BAC [III. 21]

= ∠ OCE.

Hence OC is a tangent to the ⊙ DEC [III. 32. Converse].

And since OC is a radius of the given ⊙, ∴ the two ⊙ˢ cut orthogonally [p. 240].

2. (i) The ∠ ACD = $\frac{1}{2}$ ∠ ACB [I. 8]

= ∠ APB [III. 20].

Similarly ∠ ADC = ∠ AQB.

∴ ∠ CAD = ∠ PBQ [I. 32].

(ii) Similarly ∠ CBD = ∠ PBQ.

From each of which take ∠ PBD.

∴ ∠ CBP = ∠ DBQ.

∴ ∠ BPC = ∠ BQD.

3. (i) Join BC, AD.

Then $PA^2 + PB^2 + PC^2 + PD^2 = BC^2 + AD^2$ [I. 47].

But since AB and CD are at rt. angles,

∴ the arcs BC and AD make up a semicircle [Ex. 1, p. 240].

Hence $BC^2 + AD^2 = (\text{diam.})^2$ [III. 31, and I. 47]

$= 4\,(\text{radius})^2$.

(ii) $AB^2 + CD^2 + 4OP^2$

$= PA^2 + PB^2 + 2PA \,.\, PB + PC^2 + PD^2$

$+ 2PC \,.\, PD + 4OP^2$ [II. 4]

$= 4\,(\text{radius})^2 + 4OP^2 + 2PA \,.\, PB + 2PC \,.\, PD$ [Ex. 3]

$= 4\,(\text{radius})^2 + 2\,(OP^2 + PA \,.\, PB) + 2\,(OP^2 + PC \,.\, PD)$

$= 4\,(\text{radius})^2 + 2\,(\text{radius})^2 + 2\,(\text{radius})^2$ [III. 35]

$= 8\,(\text{radius})^2$.

4. Let AC, BD be the two par[l]. tangents, O the centre of the given ⊙, and let CD be the third tangent touching the ⊙ at P. Join OP, OC, OD.

Then the ∠ COD is a rt. angle [Ex. 5, p. 235].

Hence a semicircle on DC as diam[r]. passes through O [III. 31].

And OP is perp. to DC.

∴ by the reasoning of II. 14, $DP \,.\, PC = OP^2$.

5. Let OA, OB be the two given st. lines, C and D the centres of the given ⊙[s], and P their point of contact. Then P is the middle point of CD [Hyp. and III. 12]. Also it is clear that C and D will move on st. lines par[l]. respectively to OA, OB, and at a distance from them equal to the radius of the given ⊙[s].

If these lines intersect at X, the locus of P is a ⊙ whose centre is X, and whose radius is equal to the radius of either of the given ⊙[s]. For XP = PC = PD [III. 31].

6. Let O be the centre of the ⊙, and Y the middle point of CD. Join XY, OC, OY.

Then OY is perp. to AB [III. 3 and Hyp.].

Now $XC^2 + XD^2 = 2\{CY^2 + XY^2\}$ [Ex. 24, p. 161]

$= 2\{CY^2 + OY^2 + OX^2\}$ [I. 47]

$= 2\{OC^2 + OX^2\}$

$= 2\{OA^2 + OX^2\}$

$= XA^2 + XB^2$ [II. 9].

7. Join PY, QX. [In the fig. taken PX and QY are on opposite sides of PQ.]

Then ∠ QPX = ∠ PQY [I. 29] = ∠ PXY [III. 21].

Also ∠ QPY = ∠ QXY [III. 21].

By addition, ∠ XPY = ∠ PXQ.

But ∠ PXQ is constant, since PQ is fixed.

∴ ∠ XPY is constant; ∴ arc XY is constant [III. 26]

∴ chord XY is constant [III. 29].

∴ XY touches a fixed concentric circle [Ex. 1, p. 235].

8. Let CD be the perp. and let CD meet the first ⊙ at G, and the second ⊙ at O and O′.

Then by Ex. 15, p. 234, since the ⊙ˢ are equal, the distances from D of the two points on the one ⊙ are respectively equal to the distances of the two points on the other.

Let O be the point corresponding to G.

Then O is the orthocentre [Ex. 21, p. 244], for DO = DG.

Otherwise. The ∠ AGB is the supplement of the ∠ ACB [III. 22]. And since the segments AGB, AOB are equal [Hyp. and III. 28].

∴ ∠ AGB = ∠ AOB, ∴ ∠ AOB is suppᵗ. of ∠ ACB.

∴ orthocentre is on arc AOB [Ex. 35, p. 245]. But the orthocentre is on perp. CD. ∴ orthocentre is at O.

9. Call the ⊙ˢ (i), (ii), (iii). Let (i) and (iii) intersect again at B, (ii) and (iii) at C, (i) and (ii) at D.

Then as in the second proof of the last exercise it may be shewn by means of the ⊙ˢ (i) and (iii) that the orthocentre of the △ABC lies on the arc ADB. Similarly, by means of the ⊙ˢ (ii) and (iii) the orthocentre of the △ABC lies on the arc ADC.

∴ the orthocentre is at D.

Hence of the four points A, B, C, D each is the orthocentre of the triangle formed by joining the other three [Ex. 24, p. 244].

10. Let A be the given point, and O the centre of the given ⊙. Join AO, and bisect it at X. With centre X and radius equal to one-half of the radius of the given ⊙, describe a ⊙ cutting the convex ○ᶜᵉ at Y.

Join AY and produce it to meet the concave ○ᶜᵉ again at B.

Then shall AY = YB.

This follows, because X is the middle point of the side AO, and XY is half the base OB [See Ex. 2 and 3, p. 104; and VI. 7, Cor.].

Impossible when the minimum distance from A to the ○ᶜᵉ is greater than the diameter.

11. Let O be the common centre. Draw OA any radius of outer ⊙. Bisect AO at P. On AO and AP describe semi-⊙ˢ, of which that on AP cuts the inner ⊙ at X. Join AX, and produce it to meet the semi-⊙ on AO at R. Join PX, OR.

Then ∠AXP = ∠ARO, for each is a rt. angle [III. 31].

∴ PX is parˡ. to OR; and P is middle point of OA.

∴ AX = XR [Ex. 1, p. 104].

Hence if AR is produced to meet the inner and outer ⊙ˢ at Y and B respectively,

AB = twice XY [III. 3], for OR is perp. to AB,

12. Join AA′. Then since the arcs AB′, AC′ are respectively half the arcs AC, AB, ∴ the ∠ˢ AA′B′, AA′C′ are respectively equal to half the ∠ˢ ABC, ACB.

Hence the ∠ B′A′C′ $= \frac{1}{2}$(B + C). And so for the other angles of the △A′B′C′.

Again AA′ makes with B′C′ an angle equal to that at the ⊙ᶜᵉ subtended by the sum of the arcs AB′, BA′, BC′ [Ex. 1, p. 240]; that is, an angle equal to $\frac{A}{2} + \frac{B}{2} + \frac{C}{2}$, or one rt. angle.

Hence AA′, BB′, CC′ are the perps. of the △A′B′C′.

Let A″B″C″ be the pedal △ of the △A′B′C′.

Then ∠ C′A″B″ = ∠ B′A″C″ = ∠ C′A′B′ $= \frac{1}{2}$(B + C) [Ex. 20, p. 243];

∴ ∠ B″A″C″ = 2 rt. angles − ∠ C′A″B″ − ∠ B′A″C″
= 2 rt. angles − (B + C)
= ∠ A.

13. Let ABCD be the quadˡ. Let AB, DC meet at P, and BC, AD at Q. Let the bisectors of the ∠ˢ at P and Q meet at O. Join PQ.

Then ∠ OPQ $= \frac{1}{2}$(∠ CPQ + ∠ APQ),

and ∠ OQP $= \frac{1}{2}$(∠ CQP + ∠ AQP),

∴ ∠ POQ $= \frac{1}{2}$(∠ PCQ + ∠ PAQ) [I. 32]
$= \frac{1}{2}$(∠ BCD + ∠ BAD) [I. 15]
= one rt. angle [III. 22].

14. Let ABCD be the quadrilateral whose sides AB, BC, CD, DA touch the inscribed ⊙ at X, Y, Z, V.

Let BA, CD, produced, meet at P; and DA, CB, produced, meet at Q. Bisect the ∠ˢ at P and Q by PO, QO.

Then PO is perp. to XZ, and QO to YV [Ex. 2, p. 196].

But since the fig. ABCD is cyclic, ∴ PO, QO are at rt. angles to one another [Ex. 13, p. 265].

Hence XZ and YV are perp. to one another.

15. Let ABC be a triangle of the system on the fixed base AB.

Produce AC to D, making CD equal to CB.

Then AD is of constant magnitude. Join BD cutting the bisector of the ∠ BCD at P. Then CP bisects BD at rt. angles. Required the locus of P.

Bisect AB at O. Join OP.

Then OP is one-half of AD [Ex. 3, p. 105].

That is, OP is constant; and since O is a fixed point, the locus of P is a circle, whose centre is at O, and whose radius is half AD.

16. Join AQ, and produce it to meet A′P′ at X. Join A′Q′.

Then, by hyp. and III. 31, AX, PP′, A′Q′ are par[l].

Also QQ′ and A′X are par[l].

Then $AA'^2 = AX^2 + XA'^2$ [I. 47]

$= P'P^2 + Q'Q^2$.

17. Let X, Y be the centres of the two ⊙s, P being on the ○ce of the ⊙ (X). Join AB, AX, AY, AD.

Then AX and AY are tangents [Hyp.].

∴ ∠ XAC = ∠ ADC [III. 32].

Also ∠ YAC = ∠ ABP [III. 32]

= ∠ ACD [III. 21].

Hence, by addition, ∠ XAY = ∠ ADC + ACD.

But ∠ XAY is a rt. angle [Hyp.];

hence ∠ DAC is a rt. angle [I. 32];

∴ DC is a diameter [III. 31].

18. Join PA, PB, PC; and bisect these lines at S_1, S_2, S_3. Then since the ∠s PEA, PFA are rt. angles, ∴ the four points P, F, E, A lie on a ⊙ whose diam. is PA. Hence S_1 is the centre of the ⊙ about the △ EPF.

Similarly for S_2 and S_3.

Again [Ex. 2 and 3, p. 104], S_1S_2, S_2S_3, S_3S_1 are par[l]. to AB, BC, CA, and equal to half of these lines,

∴ the △ $S_1S_2S_3$ is equiangular to the △ ABC.

19. Take O the centre of the ⊙. Join PC: then PC produced must pass through O. Join OA, OX, OY.

Then since the ∠s PAO, ACO are rt. angles,

∴ PA must touch the ⊙ circumscribed about the △ACO.

∴ $PC . PO = PA^2 = PX . PY$ [III. 36].

∴ the four points X, C, O, Y are concyclic.

∴ ∠XCP = ∠OYX [Ex. 5, p. 241]

= ∠OXY [I. 5]

= ∠OCY [III. 21].

And AC is perp. to PO; ∴ CA bisects the ∠XCY.

20. Let AB be the sum of the lines, K the side of the sq. to which the rectangle is equal.

On AB describe a semi-⊙, and draw a st. line par[l]. to AB at a distance from it equal to K, cutting the semi-⊙ at P, P′. From P (or P′) draw PX perp. to AB. Then AX, XB are the required st. lines; for $AX . XB = PX^2$.

This may be proved as in II. 14, or as a special case of III. 35.

21. Let the sum of the sqq. on required lines be equal to the sq. on AB, and the rectangle contained by them to the sq. on K.

Analysis. On AB describe a semi-⊙: then if any point P is taken on the ○ce, we shall have $AP^2 + PB^2 = AB^2$ [III. 31, I. 47].

Hence we have to find a point P on the ○ce such that AP . PB = sq. on K. Suppose PX drawn perp. to AB.

Then △APB = $\frac{1}{2}$rect. AP, PB; for ∠APB is a rt. angle.

Also △APB = $\frac{1}{2}$rect. AB, PX [I. 41].

Hence rect. AP, PB = rect. AB, PX.

Construction. To AB apply a rectangle equal to the sq. on K [I. 45].

And let D be its altitude. Draw MN par^{l}. to AB at a distance from it equal to D, cutting the semi-⊙ at P or P′. Then evidently AP, BP are the lines required.

22. Let K be the sum of the required lines, and let the sq. on AB be equal to the sum of the sqq. on them.

On AB describe a semi-circle, and also a segment containing an angle equal to half a rt. angle.

From centre A, with radius K, draw a ⊙ cutting the latter segment at D (or D′). Join AD cutting the semi-⊙ at P. Join PB. Then shall AP, BP be the required lines. Join DB.

For $AP^2 + PB^2 = AB^2$ [III. 31, and I. 47].

Also ext. ∠ APB = ∠ PDB + ∠ PBD [I. 32].

But ∠ PDB = $\frac{1}{2}$ ∠ APB; ∴ ∠ PBD = $\frac{1}{2}$ ∠ APB [Constr.].

∴ ∠ PDB = ∠ PBD; ∴ PD = PB.

∴ AP + PB = AP + PD = AD = K.

23. Let AB be the diff. of the required lines, and let the rect. contained by them be equal to the sq. on K.

On AB as diam. describe a ⊙.

At any point T on the ○^{ce} draw a tangent TP, making TP equal to K. Take the centre O, and draw PQOR cutting the ○^{ce} at P, Q. Then shall PQ, PR be the required lines.

For rect. PR, PQ = the sq. on PT [III. 36]

= the sq. on K.

And the diff. of PR and PQ is QR, that is, AB.

24. Let AB be the sum or diff. of the required lines, and the sq. on CD the diff. of the sqq. on them.

Draw DE perp. to CD, and of any length. Join CE.

Then $CD^2 = CE^2 - ED^2$ [I. 47].

From centre A, with radius CE, describe a ⊙.

From centre B, with radius DE, describe a ⊙, cutting the other ⊙ at P (or P′).

From P draw PX perp. to AB, or AB produced.

Then AX and BX shall be the required lines.

For $AX^2 + PX^2 = AP^2$, and $BX^2 + PX^2 = BP^2$ [I. 47];

$\therefore\ AX^2 - BX^2 = AP^2 - BP^2 = CE^2 - ED^2 = CD^2$.

And $AX \pm BX = AB$.

25. The $\angle PAP'$ must be a rt. angle [Ex. 2, p. 35].

So that $OP = OA = OP'$ [III. 31].

Now $\angle OAC = \angle OAP - \angle CAP$

$= \angle OPA - \angle PAB$ [I. 5 and Hyp.]

$= \angle ABC$ [I. 32],

$\therefore$ OA is a tangent to the ⊙ about the $\triangle ABC$ [III. 32].

26. Let the feet of the perps. be D, E, F.

[Take the figure, as on p. 250, in which F is in AB produced.]

(i) The four points E, C, P, D are evidently concyclic,

$\therefore\ \angle ECD = \angle EPD$ [III. 21].

That is, $\angle ACB = \angle A'PB'$.

Hence arc $A'B'$ = arc AB. Hence chord $A'B'$ = chord AB.

And so for the other sides. Hence the $\triangle^s$ are identically equal.

(ii) Join $A'B$.

Since arc AB = arc $A'B'$, $\therefore\ \angle AA'B = \angle A'BB'$ [III. 27];

$\therefore\ AA'$ is par[l]. to BB' [I. 27].

27. Take two lines of the system PQ and *pq*.

It may be proved by method similar to that of Ex. 1, p. 211, that arc P*p* = arc Q*q*. Hence the chord P*p* is equal and par[l]. to chord Q*q*. From which it follows that PQ = *pq* [I. 33].

The prop. may also be proved by noting that P is the orthocentre of the $\triangle AQB$ [Ex. 35 and Ex. 31, p. 245].

28. With figure of p. 243, let S be the centre of circum. ⊙, and let SA meet EF at X.

Then $\angle AFX = \angle ACB$ [Ex. 20, Cor. ii, p. **243**].

And $\angle ASB$ = twice $\angle ACB$ [III. 20],

$\therefore\ \angle SAB$ = comp[t]. of $\angle ACB$ [I. 5, I. 32].

Hence from $\triangle AFX$, the $\angle AXF$ is a rt. angle [I. 32].

29. Since the ∠ˢ CEP, CDP are rt. angles, the ⊙ about the △PED passes through C, and is described on PC as diam. Hence it is required to find the locus of X, the middle point of CP.

Take S the centre of the ⊙, and join SX.

Then since the ∠ CXS is a rt. angle [III. 3], and the points C, X are fixed, ∴ the locus is a ⊙ on CS as diam.

30. Take the figure of p. 250. Draw the diam. AX, and join AP, PX.

Then the four points P, D, E, C are concyclic.

∴ ∠ EDC = ∠ EPC [III. 21]
= compᵗ. of ∠ PCE
= compᵗ. of ∠ PXA [III. 21]
= ∠ PAX [III. 31].

31. Let ACD, AEF and BEC, BFD be the two pairs of lines. Let the ⊙ˢ about the △ˢ ACE, BEF meet at P. Then shall the ⊙ˢ about the △ˢ AFD, BCD pass through P. Join PF, PE, PA.

Then ∠ BFP = ∠ BEP [III. 21]
= ∠ PAC [Ex. 5, p. 241].

Hence ∠ˢ PAD, PFD together = two rt. angles.

∴ the points A, D, F, P are concyclic; that is, the ⊙ about the △ADF passes through P.

The proposition may also be proved by the properties of Simson's Line [See Exx. 74, 77, p. 250].

32. From the last exercise it is seen that the ⊙ˢ about the four △ˢ pass through a common point P. Hence it may be seen (Ex. 77, p. 250) that the four △ˢ have a common pedal for the point P.

Also (Ex. 78, p. 251) this pedal bisects each of the lines joining P to the four orthocentres.

Hence, by the method of Ex. 2, p. 124, the orthocentres are collinear.

THE CONSTRUCTION OF TRIANGLES. **Page 267.**

33. Let PAB be the given vertical angle, AB the given side, and K the given altitude.

From centre A, with radius K, describe a ⊙; and from B draw BDC to touch the ⊙ at D, and meet AP at C.

Then ABC is the required triangle.

For AD is perp. to BC (III. 18), and is equal to K.

34. Let D, E, F be the feet of the perps.

Bisect the ∠s DEF, EFD, FDE, by lines which meet at O [Ex. 2, p. 111].

Draw lines through D, E, F perp. to OD, OE, OF.

Then any one of the four △s ABC, OBC, OCA, OAB thus formed will satisfy the given conditions. [See Ex. 20, p. 243].

35. Let AB be the given base, H the given altitude, and K the radius of the circum. ⊙.

Draw PQ par[l]. to AB and at a distance from it equal to H. Then the vertex of required △ lies on PQ.

Bisect AB at rt. angles by XY; then the centre of the circum. ⊙ lies on XY [III. 1].

From centre A, with radius K, intersect XY at S.

Lastly, from centre S, with radius K, intersect PQ at C or C′. Then either of the △s ABC, ABC′ satisfies the conditions.

36. On AB, the given base, describe a segment containing the given angle; then the vertex of the required △ must be on the arc of this segment. Bisect AB at X.

Hence it is required to find a point C on this arc such that

$$AC^2 + BC^2 = K^2.$$

But $AC^2 + BC^2 = 2\{AX^2 + XC^2\}$ [Ex. 24, p. 161].

∴ $2\{AX^2 + XC^2\} = K^2.$

But AX is known, and K is given; hence XC can be determined [II. 14, I. 47].

From centre X, with radius XC describe a ⊙ cutting the segment at C, C′.

Then either of the △s ABC, ABC′ satisfies the given conditions.

37. Let AB be the given base, and H the given altitude.

Draw PQ par[l]. to AB and at a distance from it equal to H. Then the vertex of the required triangle must lie on PQ. Then apply the method of the last Exercise.

38. On AB, the given base, describe a segment of a ⊙ containing the given angle; then the vertex of the required △ must lie on its arc.

Divide AB internally or externally at X, so that $AX^2 - BX^2 =$ the given square [See *Constr.* of Ex. 24, p. 266].

From X draw XC perp. to AB to meet the arc at C. Then the △ ABC satisfies the given conditions.

For $AC^2 = AX^2 + XC^2$; and $BC^2 = BX^2 + XC^2$.

∴ $AC^2 - BC^2 = AX^2 - BX^2 =$ the given square.

[NOTE. If the side of the given square is greater than AB, X is external to AB, in which case there may be two solutions.]

39. Let H and K be the lengths of the two medians, and X the given angle. Draw DB equal to H, and bisect DB at P.

On PB draw a segment of a ⊙ containing an angle equal to X. Mark off DG one-third of DB [Ex. 19, p. 107].

And from centre G, with radius equal to one-third of K, intersect the arc at F. Join FG, and produce it to C, making GC double of GF. Join CD and BF, and produce them to meet at A. Then ABC shall be the required △.

For $DG = \frac{1}{3}DB$, and $DP = \frac{1}{2}DB$, ∴ $GP = \frac{1}{6}DB$.

∴ DG is double GP: also CG is double GF [*Constr.*].

∴ PF is par[l]. to CA [VI. 2, or by the method of Ex. 10, p. 79].

∴ ∠ CAB = ∠ PFB = ∠ X.

And since PF is par[l]. to AC, and P is the middle point of BD, ∴ F is the middle point of AB. Also AD = DC, for each is double of FP.

40. On the given base AB describe a segment containing the given angle, and another segment containing half the given angle. Make the ∠ ABK half the given diff. of the base angles: and draw BD perp. to BK to cut the larger segment at D. Join AD, cutting the smaller segment at C and BK at Q. Then ABC is the △ required.

For ∠ ACB = ∠ CDB + ∠ CBD, and ∠ CDB = $\frac{1}{2}$ ∠ ACB;

∴ ∠ CBD = $\frac{1}{2}$ ∠ ACB. ∴ ∠ CDB = ∠ CBD; ∴ CD = CB.

And KBD is a rt. angle. Hence CQ = CB [III. 31].

∴ BK is perp. to the bisector of the vertical ∠ ACB.

∴ ∠ ABK = $\frac{1}{2}$ diff. of ∠ˢ CBA, CAB [Ex. 7, p. 109].

41. Let the bisector of the vert. ∠ be a part of PQ, a line of unlimited length. Bisect AB, the given base, at rt. angles, by a line which cuts PQ at X.

Describe a ⊙ about AXB, and let it cut PQ again at C. Then the △ ABC will be that required.

For chord AX = chord BX [I. 4].

∴ arc AX = arc BX [III. 28].

∴ ∠ ACX = ∠ BCX [III. 27].

42. On AB the given base describe a segment containing the given angle. Then the vertex of the required △ must lie on the arc. Complete the ⊙.

Analysis. Let C be the vertex. Take X the middle point of the conjugate arc AB. Join XC cutting AB at D. Then DC is the bisector of the vert. ∠, for the arc AX = the arc BX.

Draw the diam. XY cutting AB at E. Then XY is perp. to AB. Join YC.

Now the points E, D, C, Y are concyclic, for the ∠ˢ YED, YCD are rt. angles.

∴ rect. XD, XC = rect. XE, XY, which is known.

Hence, given the rect. contained by XD, XC and DC the diff. between these lines, the lengths XD, XC may be found by Ex. 23, p. 266. Thus the necessary construction is obtained.

43. Draw AD equal to the given perp., and through D draw PQ perp. to AD. From centre A with radii equal to the bisector of the vert. ∠, and the median cut PQ at E and F on the same side of D [See Exx. 12, 13, p. 102].

At F draw FX perp. to PQ to cut AE produced at X. Draw AS, making the ∠ XAS equal to the ∠ AXF, and cutting XF produced at S.

From centre S, with radius SA, or SX, describe a ⊙, cutting PQ at B, C. Then ABC is the △ required.

For since SX, drawn from the centre, cuts BC at rt. angles, ∴ F is the middle point of BC; ∴ AF is the median.

Also, chord BX = chord CX [I. 4]; ∴ arc BX = arc CX [III. 28].

∴ ∠ BAX = ∠ CAX [III. 27].

44. Consult Ex. 8, p. 109, and the last example.

Let AE be the bisector of the vertical ∠; on AE describe a rt. angled triangle ADE, making the ∠ EAD half the diff^{ce}. of the base angles. Then AD is the altitude of the required △ [Ex. 8, p. 109], and ED produced is the base line.

From centre A, with radius equal to the given median, cut DE produced at F. Join AF.

Then we have given the position and magnitude of the altitude, bisector of vert. ∠, and median from A.

Hence the problem is reduced to that solved in the last exercise.

BOOK IV.

EXERCISES.

Page 269.

1. Let A be the given point without or within the given ⊙, and let K be the length of the required chord.

Place any chord PQ equal to K in the given ⊙, and describe a concentric ⊙ to touch PQ.

From A draw a tangent AT to the ⊙ of construction, cutting the given ⊙ at X, Y. Then XY is the required chord

[Proof by III. 18 and III. 14, as in Ex. 3, p. 197].

Impossible, if A is outside the given ⊙, when K is greater than the diam.

Impossible, if A is within the given ⊙, when K is greater than the diam., or when A falls within the ⊙ of construction.

2. Let O be the centre of the given ⊙, AB the given st. line, and K the length of the required chord.

Place any chord PQ equal to K in the given ⊙, and, as before, describe a concentric ⊙ to touch PQ.

From O draw a perp. to AB cutting the ⊙ of construction at C. Through C draw XCY perp. to OC cutting the given ⊙ at X, Y.

Then XY shall be the required chord.

Proof as in the last Exercise.

Page 275.

1. Let ABC be the inscribed equilat. △, and YZ, ZX, XY the tangents through the points A, B, C.

Then each of the ∠s ZAB, ZBA, = ∠ ACB [III. 32]

$= \frac{1}{3}$ two rt. angles.

∴ the ∠ Z = $\frac{1}{3}$ two rt. angles [I. 32]: and so for the ∠s X and Y, ∴ the △ XYZ is equilat. [I. 6, Cor.].

Again each of the △s XBC, YAC, ZAB may be shewn identically equal to the △ ABC [I. 26].

∴ area of △XYZ is four times △ABC.

2. Let AB, CD be the two par[l]. lines cut by PQ.

Bisect the ∠s APQ, CQP by lines which meet at O. Draw OX, OY, OZ perp. to AB, CD, PQ.

Then the △s XPO, ZPO are identically equal [I. 26].

∴ OX = OZ. Similarly OZ = OY. And since the ∠s at X, Y, Z are rt. angles, a ⊙ described from centre O with radius OX touches the given lines at X, Y, Z.

A second ⊙ is obtained by bisecting the ∠s BPQ, DQP.

Again since OX, OY are perp. to par^l. lines, they may be shewn to be in the same st. line.

Hence the diam. of each ⊙ is the perp. distance between the given par^{ls}.; ∴ the ⊙^{s} are equal.

3. For if ⊙^{s} are described about the △^{s}, the segments containing the △^{s} are on equal bases and contain equal angles, ∴ they are equal [III. 24].

Hence the two ⊙^{s} of which these segments are parts must be equal [III. 10, Cor. (ii)].

4. For because the inscribed circle touches AB and AC, ∴ its centre I lies on the bisector of the ∠ BAC [Ex. 1, p. 196].

Similarly I_1 lies on the same bisector:

∴ A, I, I_1 are collinear.

5. Let ABC be the △, and I the centre both of the circumscribed and inscribed ⊙^{s}.

Then (i) IA = IB = IC; and (ii) IA, IB, IC must be the bisectors of the ∠^{s} BAC, ABC, ACB.

Since IA = IB, ∴ ∠ IAB = ∠ IBA;

but ∠ BAC = twice ∠ IAB; and ∠ ABC = twice ∠ IBA,

∴ ∠ ABC = ∠ BAC.

Similarly ∠ ACB = ∠ BAC = ∠ ABC.

∴ △ is equilateral [I. 6, Cor.]

6. Join BS, CS.

Then since SA = SB, ∴ ∠ SAB = ∠ SBA.

And since SA = SC, ∴ ∠ SAC = ∠ SCA.

But since I is in AS, ∴ ∠ SAB = ∠ SAC.

∴ ∠ SBA = ∠ SCA.

Hence the △^{s} BAS, CAS may be shewn identically equal [I. 26].

∴ AB = AC.

7. Let ABC be the △ rt.-angled at C, and let the inscribed ⊙ touch the sides BC, CA, AB at D, E, F; and let I be its centre.

Join ID, IE. Then clearly the fig. IC is a square.

Hence the diam. of inscribed ⊙ = DC + CE.

And since BCA is a rt. angle, the diam. of the circum. ⊙ = BA [III. 31]

= BF + AF = BD + AE [III. 17. Cor.].

∴ the sum of the diams. = DC + CE + BD + AE = BC + AC.

8. In the △AFE, since AF = AE, ∴ ∠AFE = ∠AEF.

∴ each of these ∠s is acute [I. 17].

But ∠AFE = ∠FDE [III. 32].

∴ the ∠FDE is acute. So for the other two angles.

Now since AF = AE, the ∠AFE is the comp^t. of half the ∠BAC [I. 32].

Hence the ∠s of the ∠DEF are the comp^ts. of half the ∠s of the △ABC.

9. Join BI, BI_1, also CI, CI_1.

Then BI, BI_1 are respectively the internal and external bisectors of the ∠ABC,

∴ the ∠IBI_1 is a rt. angle [Ex. 2, p. 35].

Similarly the ∠ICI_1 is a rt. angle.

∴ the four points I, B, I_1, C are concyclic [III. 31].

10. Take the figure of p. 272.

Then since AG = AE [III. 17. Cor.], it follows that the diff. of AC and AB = the diff. of GC and EB, that is the diff. of CF and BF.

11. Let A, B, C denote the ∠s of the given △.

Then A + B + C = two rt. angles [I. 32],

$$\therefore \tfrac{1}{2}A + \tfrac{1}{2}B + \tfrac{1}{2}C = \text{one rt. angle},$$

$$\therefore \tfrac{1}{2}A + \tfrac{1}{2}B \text{ is the comp}^t\text{. of } \tfrac{1}{2}C.$$

But ∠BID = $\tfrac{1}{2}$A + $\tfrac{1}{2}$B, from △BAI [I. 32],

And ∠EIC = the comp^t. of $\tfrac{1}{2}$C, from △EIC

∴ ∠BID = ∠EIC.

12. [In the figure taken AB is greater than AC.]

Since the $\angle$ ASB is twice the $\angle$ ACB [III. 20],

$\therefore \angle$ SAB = one rt. angle – C [I. 32],

$$\begin{aligned}\therefore \angle SAI &= \angle IAB - \angle SAB\\ &= \tfrac{1}{2}A - (\text{one rt. angle} - C)\\ &= \tfrac{1}{2}A - (\tfrac{1}{2}A + \tfrac{1}{2}B + \tfrac{1}{2}C) + C\\ &= \tfrac{1}{2}C - \tfrac{1}{2}B.\end{aligned}$$

13. See the last Exercise, and Ex. 8, p. 109.

14. Take the figure of p. 272. Join AI.

The area of the $\triangle$ABC = the sum of the areas of the $\triangle$s IBC, ICA, IAB = the sum of half the rects. contained by BC, IF, by CA, IG and by AB, IE [I. 41] = the rectangle contained by the radius and half the sum of the sides [II. 1].

15. Let P, Q, R, S be the centres of the $\odot$s about the $\triangle$s in the order named.

Then clearly SR bisects DO at rt. angles [IV. 5].

Similarly PQ bisects OB at rt. angles.

$\therefore$ SR is par[l]. to PQ. Thus also SP is par[l]. to RQ;

$\therefore$ SPQR is a par[m].

16. Join BO, BI, OC.

Since $\angle$ BAO = $\angle$ CAO, $\therefore$ arc BO = arc OC [III. 26],

$\therefore$ chord BO = chord OC.

Again the ext. $\angle BIO = \angle IAB + \angle IBA = \tfrac{1}{2}A + \tfrac{1}{2}B$.

Also the $\angle IBO = \angle OBC + \angle IBC = \angle OAC + \angle IBC$ [III. 21]

$= \tfrac{1}{2}A + \tfrac{1}{2}B$.

$\therefore \angle BIO = \angle IBO$, $\therefore$ OI = OB = OC.

$\therefore$ O is the centre of the $\odot$ about BIC.

17. Let AB be the given base. Draw PQ par[l]. to AB at a distance from it equal to the given altitude.

Describe $\odot$s from centres A and B with radius equal to the given radius of circum.$\odot$; let these $\odot$s intersect at O, on the same side of AB as PQ. From centre O with radius OA describe a $\odot$ cutting PQ at C or C′. Then either of the $\triangle$s ABC, ABC′ satisfies the required conditions.

18. Find I the centre of the ⊙ inscribed in the △ ABC formed by the three given st. lines.

From centre I, with any radius greater than that of the inscribed ⊙, describe a ⊙; this will intercept equal chords from the sides of the △, because the perps. from the centre on these chords are equal, being radii of the inscribed ⊙.

19. Let ABC be an equilat. △, I the centre of the inscribed ⊙, I_1 the centre of the escribed ⊙ touching BC. Then I is also the centre of the circum. ⊙ and the intersection of the medians. Let AII_1 cut BC at D.

Then ID, IA, I_1D are the radii of the inscribed, circumscribed and escribed ⊙s. And IA = twice ID [Ex. 4, p. 113].

Also ∠ ABD = ∠ $I_1BD = \frac{1}{3}$ of two rt. angles;

hence △s I_1BD, ABD are identically equal [I. 26].

∴ I_1D = AD = three times ID [Ex. 4, p. 113].

20. Then AB, BC, CA pass through F, D, E [III. 12].

And the common tangents at F, D, E meet at a point O, and are equal [Ex. 16, p. 236]. Hence O is the centre of the ⊙ about EDF.

Again, since O is the intersection of tangents at E and F, ∴ O lies on the bisector of the ∠ A. Similarly O is on the bisectors of the ∠s B and C: also OF, OE, OD are perp. to the sides of the △ ABC [III. 18]. Hence the ⊙ EFD is inscribed in the △ ABC.

Page 281.

1. See solution of Ex. 7, p. 235.

2. See solution of Ex. 8, p. 235.

3. For the sum of one pair of opp. sides of a quadrilat. about a ⊙ = the sum of the other pair. But if the quadrilat. is a par^{m}., the opp. sides are equal; hence the figure must be equilateral, that is, either a rhombus or a square.

4. For if a quadrilat. is cyclic, the opp. angles together = two rt. angles [III. 22]: and if the quadrilat. is a par^{m}., the opp. ∠s are equal, ∴ each is a rt. angle.

5. See solution of Ex. 17, p. 263.

6. Take AC, BD any two diams., and draw perps. to them at their extremities, the resulting figure is circumscribed about the ⊙ [III. 16], and may be proved to be a rhombus by Ex. 3.

7. For the sides of all such squares are equal to the diameter of the ⊙.

8. In the figure of p. 278, join AB, BC, CD, DA.

Then ABCD is the inscribed square [IV. 9].

And the sqq. GE, AD, BC, EK are respectively double of the △s BEA, AED, BEC, CED. ∴ the whole fig. GK is double of the sq. ABCD.

9. The angle subtended by a side of an inscribed square at any point on the major arc is half the angle subtended at the centre, that is, half a right angle.

But the sum of the angles in the major and minor arc is two rt. angles [III. 22],
hence the angle in the minor arc is $\frac{3}{2}$ of a rt. angle.

10. In BC, CD, DA make BY, CZ, DW each equal to AX.

Join XY, YZ, ZW, WX. Then XYZW shall be the sq. required.

For the △s XBY, YCZ, ZDW, WAX are all identically equal [I. 4], ∴ the fig. XYZW is equilateral.

Also, ∠ ZYC = ∠ YXB;

∴ ∠s ZYC, XYB = ∠s YXB, XYB = one rt. angle [I. 32].

∴ ∠ XYZ is a rt. angle. Similarly each of the other ∠s of the figure is a rt. angle: ∴ it is a square.

11. The sq. of minimum area is that obtained by joining in order the middle points of the sides of the given square.

For a square is a minimum when its diagonal is a minimum: and the least line that can be drawn between two opp. sides of the given square is perp. to those sides: this is obtained by joining the middle points.

12. (i) The intersection of the diagonals is the centre [IV. 9].

(ii) On AB, CD, two opp. sides of the rect., as hypotenuse, describe two right-angled isosceles △s AXB, CYD externally to the rectangle. XA, XB, YC, YD produced will form the required square.

13. (i) Let OAB be the quadrant, AB being the arc.

Bisect the rt. angle AOB by OD; draw DF perp. to OA; bisect the ∠ ODF by DE; and at E in OA draw EC perp. to OA, meeting OD in C. Then shall C be the centre of the required ⊙.

For ∠ CED = alt. ∠ EDF = ∠ EDC [*Constr.*].

∴ CD = CE. And if CG is drawn perp. to OB, then CE = CG, for the fig. GE is a square.

Finally, since the ∠s at E and G are rt. angles, and since C is in OD, ∴ a ⊙ described from centre C with radius CD touches the arc and the radii of the quadrant.

(ii) In this question it is understood that one angle of the square is to coincide with the angle between the radii.

Bisect the ∠ AOB by OD, and draw DF, DH perp. to OA, OB.

Then OFDH is the square required.

For ∠ FOD = $\frac{1}{2}$ rt. angle, and ∠ HOF is a rt. angle.

∴ ∠ ODF = $\frac{1}{2}$ rt. angle; ∴ OF = DF. And since the fig. is a rectangular par^{m}., it is a square.

(If two angular points of the sq. are to lie on the arc of the quadrant, and the other two on the bounding radii, see Ex. 3, p. 390.)

14. Join AC, BD, and let O be the centre. Join PO.

Then in the △ APC

$$PA^2 + PC^2 = 2PO^2 + 2AO^2 \text{ [Ex. 24, p. 161]}$$
$$= 4\,(\text{radius})^2 = (\text{diam.})^2.$$

Similarly in △ BPD,

$$PB^2 + PD^2 = 2PO^2 + 2BO^2$$
$$= 4\,(\text{radius})^2 = (\text{diam.})^2.$$

∴ $PA^2 + PB^2 + PC^2 + PD^2$ = twice sq. on diam.

Page 283.

1. For since each of the base angles is double of the vert. angle, ∴ the sum of the ∠ˢ = 5 times the vert. ∠.

That is, 5 times the vert. ∠ = 2 rt. angles [I. 32].

∴ the vert. ∠ = $\frac{1}{5}$ of 2 rt. angles.

2. Describe an isosceles △ ABD, having each of the ∠ˢ at the base double of the vert. ∠ [IV. 10].

Bisect the vert. ∠ BAD, by a st. line AX: then each of the ∠ˢ BAX, DAX is one-fifth of a right angle.

Hence a rt. angle may be divided into five equal parts.

3. The △ACD in the figure of p. 282 is of the kind required. For the ext. ∠ ACD = ∠ ABD + ∠ BDC [I. 32];

and ∠ ABD is double of ∠ CDA or of ∠ CAD [IV. 10].

∴ ∠ ACD is three times either of the ∠ˢ CDA or CAD.

4. For ∠ ADB = ∠ AFD [III. 32].

And since AD = AF (radii), ∴ ∠ ADF = AFD.

Hence the two △ˢ ABD, ADF have two angles of one equal to two angles of the other, and the side AD common, ∴ BD = DF.

5. For these two circles circumscribe △ˢ which have equal bases BD, CD, and equal vert. ∠ˢ BAD, CAD

[See Ex. 3, p. 275].

6. (i) For BD, DF are equal chords [Ex. 4] subtending at the centre of the ⊙ in which they are placed angles equal to $\frac{1}{5}$ of two rt. angles [Ex. 1], that is, $\frac{1}{10}$ of four rt. angles.

(ii) For CD subtends at the ○ᶜᵉ of the ⊙ ACD an angle equal to $\frac{1}{10}$ of four rt. angles.

∴ CD subtends at the *centre* of the ⊙ ACD an angle equal to $\frac{1}{5}$ of four rt. angles.

7. Take S′ the middle point of the arc CD.

Join AS′, S′D.

Then because the arc CS′ = the arc S′D, ∴ AS′ bisects the ∠ BAD [III. 27].

But ∠ CAS′ = ∠ CDS′ [III. 21], and the whole ∠ BAD = the whole ∠ CDB [IV. 10], ∴ DS′ also bisects the ∠ CDB.

And since the $\triangle^s$ BAD, BDC are both isosceles, ∴ the bisectors of the vert. $\angle^s$ also bisect the base at rt. angles [I. 4].

∴ AS′ and DS′ if produced, would bisect BD and BC at rt. angles.

∴ S′ is the centre of the ⊙ about the $\triangle$BDC.

8. Now by the last exercise, the bisectors of the $\angle^s$ BAD, BDC meet at S′ the centre of the ⊙ about the $\triangle$BDC. If AS′ meets DC at I, then I is the centre of the ⊙ inscribed in the $\triangle$ABD, for DC bisects the $\angle$ADB. Again, if BI meets DS′ at I′, then I′ is the centre of the ⊙ inscribed in the $\triangle$DBC.

And the ext. $\angle$ S′II′ = $\angle$ IBA + $\angle$ IAB [I. 32]

$= \frac{3}{2}$ of vert. $\angle$ BAD.

Also the ext. $\angle$ S′I′I = $\angle$ IBD + $\angle$ BDI′

$= \frac{3}{2}$ of vert. $\angle$ BAD.

∴ $\angle$ S′II′ = $\angle$ S′I′I : ∴ S′I = S′I′.

Page 288.

1. Take O the centre of the ⊙; join AO cutting the ○^{ce} at D. At D draw the tangent, cutting AB, AC at X and Y. A circle inscribed in the $\triangle$AXY will touch AB, AC and the convex arc BC.

2. For the $\angle$ ABC = the $\angle$ ACB [Hyp.]

= the $\angle$ BED [III. 21].

Hence by the converse of III. 32, AB touches the ⊙ about the $\triangle$EBD.

3. The sq. inscribed in a ⊙ is clearly twice the sq. on the radius [I. 47].

Let ABC be an equilat. $\triangle$ inscribed in the same ⊙, of which O is the centre. Join AO, BO, and produce AO to meet BC at D.

Then since the triangle is equilateral, O is both the intersection of the medians and of the perps.

Hence $AB^2 = AO^2 + OB^2 + 2AO . OD$ [II. 12]

$= AO^2 + OB^2 + OA^2$, for $AO = 2OD$ [Ex. 4, p. 113]

$= 3$ times sq. on radius.

∴ twice sq. on AB = three times sq. inscribed in the ⊙.

4. Bisect the ∠ ACB by CD: then BC is a tangent to the ⊙ about the △ADC; and DA = DC = BC [IV. 10].

Now $AB^2 = AB . BD + AB . AD$ [II. 2]

$= BC^2 + AB . BC$ [III. 36].

Page 293.

1. Apply I. 32, Cor. 1, remembering that in a regular polygon the interior angles are equal.

(i) *Pentagon.* The five int. ∠ s + 4 rt. ∠ s = 10 rt. ∠ s,

∴ one int. ∠ = $\frac{6}{5}$ rt. ∠.

(ii) *Hexagon.* The six int. ∠ s + 4 rt. ∠ s = 12 rt. ∠ s,

∴ one int. ∠ = $\frac{4}{3}$ rt. ∠.

(iii) *Octagon.* Each int. ∠ = $\frac{3}{2}$ rt. ∠.

(iv) *Decagon.* Each int. ∠ = $\frac{8}{5}$ rt. ∠.

(v) *Quindecagon.* Each int. ∠ = $\frac{26}{15}$ rt. ∠.

2. Circumscribe a circle about the given pentagon [IV. 14].

Then in the figure of p. 284.

Since the chords BC, CD, DE are all equal,

∴ the minor arcs BC, CD, DE are all equal [III. 28].

∴ the ∠ s at the ○ce BAC, CAD, DAE are all equal [III. 27].

3. Solved by the method of the last Exercise.

4. (i) *Pentagon.* See figure of p. 284.

Let CD be the given base. Draw an isosceles △FGH, having each of the base ∠ s double of the vert. ∠; and on CD as base describe a △ACD equiangular to the △FGH.

About the △ACD describe a ⊙, then proceed as in IV. 11.

(ii) *Hexagon.* [See figure of p. 290.]

Let AB be the given base. On AB describe an equilat. △ ABG. From centre G with radius GA describe a ⊙, which will pass through B. Then proceed as in IV. 15.

(iii) *Octagon.* Let AB be the given base; produce AB to X, and make the ∠ XBD half a rt. angle.

Make BD equal to AB. Through the points A, B, D describe a ⊙. This ⊙ circumscribes the required octagon. Hence proceed as indicated in Note, p. 295.

5. Let ABCDEF be a regular hexagon inscribed in a circle. Then AEC may be shewn [I. 4] to be an equilat. △. Take O the centre. Join OA, OE, OC.

(i) Then the △s AFE, AOE are identically equal [IV. 15 and I. 8].

Similarly the △s ABC, AOC, and the △s EDC, EOC.

Hence the hexagon is double the equilat. △.

(ii) Let AO produced meet EC at X. Then since the △ AEC is equilat., AX is perp. to EC, and O is the intersection of medians: ∴ AO = twice OX [Ex. 4, p. 113].

And $AE^2 = AO^2 + EO^2 + 2AO . OX$ [II. 12]

$= AO^2 + EO^2 + AO^2 = 3\ (\text{radius})^2$

= three times sq. on the side of the hexagon.

6. (i) For by Ex. 2, p. 293, we have

∠ HAB = ∠ HBA = ∠ BCH,

each being one-third of the angle of a regular pentagon.

Hence the ∠ CBH is two-thirds of the angle of the pentagon.

Also the ext. ∠ CHB = ∠ HAB + ∠ HBA [I. 32]

= two-thirds of the ∠ of the pentagon.

∴ ∠ CHB = ∠ CBH; ∴ CH = CB = BA.

Similarly HE = AB.

(ii) And since ∠ ABH = ∠ BCH,

∴ AB touches the ⊙ about △ BHC [*Converse*, III. 32].

(iii) Hence AC . AH = AB^2 [III. 36]

= CH^2,

or AC is divided in medial section [II. 11].

7. Let ABCDE be a regular pentagon, and let AC, AD cut BE at P and Q. Call the interior figure PQRST.

Then since each of the ∠ˢ PAB, PBA is $\frac{1}{3}$ of an int. ∠ of the pentagon;

∴ AP = BP.

And since each of the ∠ˢ APQ, AQP is $\frac{2}{3}$ of an ∠ of the pentagon [I. 32],

∴ AP = AQ.

∴ all lines of the type AP, AQ, &c. are equal.

And all ∠ˢ of the type PAQ, QER, &c. are equal,

∴ all bases such as PQ, QR, &c. are equal.

And since the △ APQ is isosceles, ∴ ∠ TPQ = ∠ RQP [I. 5].

Hence the fig. PQRST is equilat. and equiangular.

8. (i) Proved by the method of the last Exercise.

(ii) Let ABCDEF denote the original hexagon. Let FB cut AE, AC in P, Q: call the interior figure PQRSTV.

First shew that all △ˢ of the type APQ are equilat.

Hence that FP = PQ = QB, &c.; so that the △ˢ AFP, APQ, AQB, &c. have equal area [I. 38].

Now the figure PQRSTV is made up of *six* equilateral △ˢ each equal to the △ APQ.

And the figure ABCDEF is made up of *eighteen* △ˢ (not all equilat.) each equal to the △ APQ.

9. The triangle formed by joining BCD in the figure of p. 292 satisfies these conditions.

For of such parts as the whole $\bigcirc^{ce}$ contains fifteen, the arc BC contains *two*, the arc CD *five*, and the arc DAB *eight*.

10. Let A, B be consecutive vertices of the inscribed hexagon: let the tangents at A, B meet at P; and let O be the centre of the $\odot$.

It is sufficient to compare the fig. OAPB and the $\triangle$ OAB.

Join OP, cutting AB at rt. angles at X [Ex. 2, p. 196].

The $\angle^s$ APX, PAX may be shewn to be $\frac{2}{3}$ and $\frac{1}{3}$ of a rt. $\angle$ respectively,

$\therefore$ it may be proved that PX $= \frac{1}{2}$ AP; and similarly AP $= \frac{1}{2}$ PO;

so that PX $= \frac{1}{4}$ PO; hence PX $= \frac{1}{3}$ OX.

Hence $\triangle$ PAB $= \frac{1}{3}\triangle$ OAB, being on the same base.

$\therefore \triangle$ OAB $= \frac{3}{4}$ fig. OAPB.

$\therefore$ inscribed hexagon $= \frac{3}{4}$ circumscribed hexagon.

THEOREMS AND EXAMPLES ON BOOK IV.

I. On the Triangle and its Circles.

Page 297.

1. (i) AE = AF, BE = BD, and CF = CD [III. 17, Cor.],

$\therefore$ AE + BD + DC $= s$,

or AE $+ a = s$, $\therefore$ AE $= s - a$.

Similarly BD $= s - b$, and CD $= s - c$.

(ii) $AE_1 = AF_1$ [III. 17, Cor.].

And $$AE_1 + AF_1 = AC + CE_1 + AB + BF_1$$
$$= AC + CD_1 + AB + BD_1$$
$$= AC + BC + AB = 2s,$$
$$\therefore\ AE_1 = AF_1 = s.$$

(iii) $CD_1 = CE_1 = AE_1 - AC = s - b$ [by (ii)]

$BD_1 = BF_1 = AF_1 - AB = s - c.$

(iv) $CD = BD_1$, for each $= s - c$ [by (i) and (iii)],

$BD = CD_1$, for each $= s - b$.

(v) $EE_1 = AE_1 - AE = s - (s - a) = a.$

Similarly $FF_1 = a.$

(vi) $$\triangle ABC = \text{the } \triangle\ BIC + \triangle\ CIA + \triangle\ AIB$$
$$= \tfrac{1}{2}ra + \tfrac{1}{2}rb + \tfrac{1}{2}rc \text{ [I. 41]}$$
$$= \tfrac{1}{2}r(a + b + c)$$
$$= rs.$$

Again, $$\triangle ABC = \triangle ABI_1 + \triangle ACI_1 - \triangle BCI_1$$
$$= \tfrac{1}{2}r_1c + \tfrac{1}{2}r_1b - \tfrac{1}{2}r_1a \text{ [I. 41]}$$
$$= \tfrac{1}{2}r_1(c + b - a)$$
$$= r_1(s - a).$$

2. (i) The points A, I, I_1, are collinear, for I and I_1 both lie on the bisector of the vert. ∠ BAC. [See p. 273 Note.]

Similarly I, I_2 and I, I_3 are on the bisectors of the ∠s ABC, ACB.

(ii) For AI_2 and AI_3 being the bisectors of opp. vert. ∠s [p. 273] are in the same st. line. So for the two other ranges.

(iii) The $\angle CAI_2 = \angle BAI_3$, being halves of opp. vert. $\angle^s$, and $\angle CAI = \angle BAI$ [Constr.].

$\therefore$ I_1IA is perp. to I_2I_3; similarly I_2IB is perp. to I_3I_1, and I_3IC to I_1I_2.

$\therefore$ I is the orthocentre of the $\triangle I_1I_2I_3$, and ABC is the pedal $\triangle$ of the $\triangle I_1I_2I_3$.

Hence the $\triangle^s$ BI_1C, CI_2A, AI_3B are equiangular [Ex. 20, Cor. ii., p. 243].

(iv) The chord of contact of the tangents to inscribed $\odot$ from A is perp. to AI [Ex. 2, p. 196], and is therefore par^l. to I_2I_3. So for the other sides. Hence the $\triangle^s$ are equiangular.

(v) and (vi) follow from the fact that I is the orthocentre of the $\triangle I_1I_2I_3$ [Exx. 24, 25, p. 244].

3. See the fig. to p. 297, and suppose the $\angle$ at C to be a rt. angle. Then the fig. IDCE is a square, and

$$r = ID = CE = s - c \text{ [Ex. 1 (i), p. 297]}.$$

Again, if the $\angle$ at C is a rt. angle, the fig. $I_1D_1CE_1$ is a square, so that $r_1 = I_1D_1 = D_1C = s - b$.

Similarly $$r_2 = s - a.$$

Lastly, let the third escribed $\odot$ touch CA produced and CB produced at E_3, D_3.

Then the figure $I_3E_3CD_3$ is a square;

and $r_3 = I_3D_3 = CE_3 = s$ [Ex. 1 (ii), p. 297].

4. (i) $DD_2 = BD_2 - BD = s - (s - b) = b$ [Ex. 1, p. 297],

$D_1D_3 = CD_3 - CD_1 = s - (s - b) = b$.

(ii) Similarly $DD_3 = D_1D_2 = c$.

(iii) $D_2D_3 = BD_2 + BD_3 = s + s - a = b + c$.

(iv) $DD_1 = BD - BD_1 = s - b - (s - c) = c - b$.

5. Follows directly from Ex. 20, p. 243, and Ex. 22, p. 244.

6. Worked out on p. 246; since the centre of the inscribed ⊙ is at the intersection of the bisectors of the angles. [Observe that the loci in Ex. 6 and Ex. 7 are parts of the *same circle*, of which II_1 is a diam., the centre being on the ○^{ce} of the circumscribed circle. Ex. 15, p. 299].

7. Take the figure of p. 298.

Since BI, BI_1 are respectively the internal and external bisectors of the ∠ ABC, ∴ ∠ IBI_1 is a rt. angle.

So also the ∠ ICI_1 is a rt. angle: ∴ the points I, B, I_1, C are concyclic.

∴ ∠ BI_1I = ∠ $BCI = \frac{1}{2}C$, and ∠ II_1C = ∠ $IBC = \frac{1}{2}B$ [III. 21].

Hence by addition the ∠ $BI_1C = \frac{1}{2}B + \frac{1}{2}C$

$= \text{comp}^t.$ of $\frac{1}{2}A$.

That is, the ∠ BI_1C is constant: and the base BC is fixed, ∴ the locus is the arc of a segment.

8. Given the base BC and the vert. angle, the vertex A must lie on the arc of a certain segment described on BC as base. That is, the three points A, B, C lie, for all positions of A, on a fixed circle; for if an arc of a circle is fixed, the whole circle is fixed [III. 10, Corollaries].

9. Take the figure of p. 298. Required the locus of I_2.

Since the ∠^{s} ICI_2, IAI_2 are rt. angles,

∴ the points I, C, A, I_2 are concyclic.

∴ ∠ II_2C = ∠ $IAC = \frac{1}{2}A$.

Hence the locus of I_2 is the arc of a segment on BC as base, capable of containing an angle equal to $\frac{1}{2}A$.

10. Let BC be the given base, X the given angle, and K the radius of the inscribed circle.

On BC describe a segment of a circle containing an angle equal to *one rt. angle* + $\frac{1}{2}X$.

Then the centre of the inscribed ⊙ must be on this arc [Ex. 36, p. 246].

Draw PQ par^l. to BC, and at a distance from it equal to K (on the same side of BC as the segment).

Then the centre of the inscribed ⊙ must be on PQ.

Hence if PQ cut the segment at I (or I′), I is the centre of the inscribed ⊙.

From centre I, with radius K, draw the inscribed ⊙, to which draw tangents from B and C. If these tangents produced meet at A, then ABC is the required triangle.

Prove by a method converse to Ex. 36, p. 246 that the

$$\angle \text{BAC} = \text{the angle X}.$$

11. Let BC be the given base, and X the given angle.

On BC describe a segment of a circle, (i) capable of containing an angle equal to *one rt. angle* $-\frac{1}{2}$X [see Ex. 7. p. 299], (ii) capable of containing an angle equal to $\frac{1}{2}$X [See Ex. 9, p. 299]. Then proceed as in the last Example.

12. On the base BC describe a segment containing *one rt. angle* + $\frac{1}{2}$ *the given angle;* then the centre of the inscribed ⊙ is on this arc [Ex. 36, p. 246].

At D, the given point in BC, draw a line perp. to BC cutting the arc at I. Then I must be the centre of the inscribed ⊙, and ID is its radius.

From this point proceed as in Ex. 10, p. 299.

13. On BC the given base describe a segment containing *one rt. angle* $-\frac{1}{2}$ *the given angle* [Ex. 7, p. 299]; then the centre of the escribed ⊙ must be on this arc. From this point proceed as in the last Example.

14. The △^{s} BAI, CAI are identically equal [I. 4];

$$\therefore \text{IB} = \text{IC}; \quad \therefore \angle \text{IBC} = \angle \text{ICB}.$$

But ∠ ABI = ∠ ICB [III. 32] = ∠ IBC.

That is, BI bisects the ∠ ABC.

∴ I is the centre of the inscribed ⊙.

Similarly, if AB, AC are produced to X and Y, BI_1, CI_1 may be shewn to be the bisectors of the ∠^{s} XBC, YCB : ∴ I_1 is the centre of an escribed ⊙.

15. Now I and I_1 lie on the bisector of the ∠ BAC.

Let AII_1 cut the circum-⊙ at P. Join PB.

Then ∠ PBI = ∠ PBC + ∠ CBI = ∠ PAC + ∠ CBI [III. 21]

$= \frac{1}{2}A + \frac{1}{2}B$.

Also, ext. ∠ PIB = ∠ IAB + ∠ IBA = $\frac{1}{2}A + \frac{1}{2}B$.

∴ ∠ PBI = ∠ PIB ; ∴ PI = PB.

And ∠ I_1BI is a rt. angle [Ex. 2, p. 29].

∴ ∠ PBI_1 = ∠ PI_1B [I. 32]. ∴ PB = PI_1.

Hence P is the middle point of II_1.

16. Take the figure of p. 298.

Since the ∠ s I_2BI_3, I_2CI_3 are rt. angles [Ex. 2 (v), p. 298],

∴ a ⊙ on I_2I_3 as diam. passes through B and C.

Let the circum-⊙ cut I_2I_3 at Q (and let Q be in AI_3). Join QB.

Then ∠ QI_3B = ∠ I_1I_3C + ∠ I_2I_3C

$= \frac{1}{2}A + \frac{1}{2}B$ [III. 21].

And ∠ I_3QB = ∠ C [Ex. 5, p. 241];

∴, from △ I_3QB, the ∠ $I_3BQ = \frac{1}{2}A + \frac{1}{2}B$ [I. 32].

∴ QI_3 = QB. Similarly QI_2 = QB.

∴ Q is the centre of the ⊙ through I_2, C, B, I_3.

[NOTE. Observe that the points P and Q in Exx. 15, 16 are the extremities of the diam. of the circum-⊙ perp. to BC].

17. In the △ formed by joining the three points A, B, C inscribe a circle: and let the points of contact be D, E, F (D being opp. to A, &c.).

Then AE = AF, BD = BF, and CE = CD [III. 17, Cor.].

Hence ⊙s described from the centres A, B, C, with radii AF, BD, CE will clearly satisfy the given conditions. There will be *four* solutions in all; for solutions may also be obtained from the three *escribed* ⊙s.

18. If DE does not touch the ⊙, from D draw DE′ to touch it, and meet AC at E′.

Then (i) DE′ = DB + E′C [III. 17, Cor.],

and (ii) DE = DB + EC [Hyp.].

Then DE′ ~ DE = EE′;

that is, DE′ = DE + EE′, or DE = DE′ + EE′;

which is impossible [I. 20].

Hence no line through D but DE does touch the ⊙.

19. The fixed circle is the escribed circle touching the side opp. to the fixed angle.

[Take the figure of p. 297]. Since $AE_1 = AF_1$ = half the perimeter [Ex. 1 (ii), p. 297], ∴ if the perimeter is given, the points E_1, F_1 are given, ∴ the escribed circle is given, which is necessarily touched by the base BC.

20. 21. It has been proved in Ex. 2, p. 298, that if I, I_1, I_2, I_3 are the centres of the inscribed and escribed circles of the △ABC, each of these four points is the orthocentre of the triangle formed by the other three, and that the original △ABC is the pedal triangle.

Hence given any three of the points I, I_1, I_2, I_3, we have only to draw the pedal triangle of the triangle so formed.

22. Let AX, AY, st. lines of unlimited length, contain the given vert. angle.

Mark off AE_1, AF_1 each equal to half the given perimeter.

Draw E_1I_1, F_1I_1 perp. to AX, AY, and prove that $I_1E_1 = I_1F_1$.

From centre I_1 describe a ⊙ touching AX, AY at E_1 and F_1.

Bisect the ∠XAY by AP, making AP equal to the given bisector. From P draw a tangent to the ⊙, meeting AX, AY at B, C. Then ABC is the triangle required; for it has the given vert. ∠, and the given bisector; and since the ⊙ E_1F_1 is an escribed ⊙, and $AE_1 = AF_1$ = the semi-perimeter [Ex. 1 (ii), p. 297], ∴ the △ABC has also the given perimeter.

23. Let AX, AY make the given vert. angle. Mark off AE_1, AF_1 each equal to half the given perimeter, and describe a ⊙ to touch AX, AY at E_1 and F_1. Then prove as in Ex. 19 that this is an escribed ⊙ of the required triangle.

From centre A, with radius equal to the given altitude, describe a ⊙. Draw either of the transverse common tangents to the two ⊙s [Ex. 17, p. 236]. If the common tangent meets AX, AY at B and C, then ABC satisfies the given conditions.

24. Let AX, AY determine the given vert. ∠. Mark off AE_1, AF_1 equal to half the given perimeter: and as in the last two examples describe a circle to touch AX, AY at E_1 and F_1.

Draw a ⊙, with radius equal to the given radius, to touch AX, AY [Ex. 32, p. 239].

And draw either of the transverse common tangents to the two ⊙s. If the common tangent meets AX, AY at B and C, then ABC is the required triangle.

25. Draw the inscribed ⊙ as in the last Example; and from the given vertex as centre, with the given altitude as radius, describe a circle. Then draw either of the direct common tangents to the two ⊙s.

26. Let BC be the given base, and K the given difference of the sides. Cut off BX equal to K, and bisect XC at D.

At D draw DI perp. to BC and equal to the given radius.

From centre I and with radius ID describe a ⊙, to which draw tangents from B and C. If these tangents intersect at A, then ABC is the required △.

For $\quad AB \sim AC = BD \sim DC$ [Ex. 10, p. 276]

$\quad\quad = BD \sim DX = BX = K.$

27. Let A be the vertex, S the centre of the circum-⊙, and I the centre of the inscribed ⊙.

From centre S with radius SA describe the circum-⊙.

Join AI, and produce it to meet the $\bigcirc^{ce}$ at X.

From centre X with radius XI cut the circum-⊙ at B and C. Then ABC shall be the required △.

Join XB, XC, BI, CI.

For, since BX = XC, ∴ ∠ BAX = ∠ CAX [III. 28, 27]; ∴ AI is the bisector of the ∠ BAC.

Again, the ext. ∠ XIC = the ∠ˢ IAC, ICA.

But ∠ XIC = ∠ XCI (since XC = XI)

= ∠ˢ XCB, BCI,

∴ ∠ˢ IAC, ICA = ∠ˢ XCB, BCI.

But ∠ IAC = ∠ XCB [III. 21], ∴ ∠ ICA = ∠ BCI.

∴ CI is the bisector of the ∠ BCA. Hence I is the centre of the inscribed ⊙.

28. Take the figure of p. 298.

Since BI and BI_1 are the internal and external bisectors of the ∠ ABC, ∴ the ∠ IBI_1 is a rt. angle.

Similarly, the ∠ ICI_1 is a rt. angle. ∴ II_1 is the diameter of the ⊙ about IBI_1C. But the ○ᶜᵉ of the ⊙ about the △ABC bisects II_1 [Ex. 15, p. 299: that is, the centre of the ⊙ about BIC lies on the ○ᶜᵉ of the ⊙ about ABC.

29. Take the figure of p. 298.

Join ID, and produce it to P, making DP equal to I_1D_1 (i.e. r_1).

Then remembering that $BD_1 = CD$ and $D_1C = BD$ [Ex. 2] we may prove that the △ˢ I_1D_1B, PDC are equal, also the △ˢ I_1D_1C, PDB are equal [I. 4]. Hence the ∠ BI_1C = ∠ BPC. ∴ P is on the ⊙ IBI_1C.

$$\therefore \text{BD . DC} = \text{PD . DI} \ [\text{III. } 35] = r_1 . r.$$

30. For the ∠ FDB = the ∠ FED [III. 32],

also the ∠ DE′F′ = the ∠ FED [Ex. 20, Cor. ii, p. 243].

∴ ∠ E′DB = ∠ DE′F′ : ∴ E′F′ is par^l. to BC [I. 27].

31. Let the inscribed ⊙ of the △ABC touch AC, AB at E and F:

and let the ⊙ˢ inscribed in the △ˢ ABD, ACD touch AD at Q and Q′.

Then by Ex. 1, p. 297, $AQ = \frac{1}{2}\{AD + AB - BD\}$

$$= \frac{1}{2}\{AD + AB - BF\}$$

$$= \frac{1}{2}\{AD + AF\}.$$

Similarly $AQ' = \frac{1}{2}\{AD + AE\}$.

But $AF = AE$ [III. 17, Cor.], $\therefore AQ = AQ'$.

That is, the two ⊙ˢ touch AD at the same point, and therefore touch one another.

ON THE NINE-POINT CIRCLE. **Page 303.**

34. Take the figure of p. 302.

If the base and vert. ∠ are given, then the circum-⊙ is fixed in position and magnitude [III. 21]; hence the radius of the nine-points ⊙ (being half that of the circum-⊙) is given: that is, XN is constant. But X is a fixed point; ∴ the locus of N is a ⊙, of which X is the centre.

35. For by Ex. 24, p. 244, each of the △ˢ ABC, AOB, BOC, COA have the same pedal triangle, and therefore the same nine-points ⊙, for the nine-points ⊙ circumscribes the pedal triangle.

36. For by Ex. 2 (v), p. 298, the △ABC is the pedal triangle of each of the four triangles formed by joining three of the points I, I_1, I_2, I_3.

37. For, in the figure of p. 302, both O and S are fixed: ∴ N, the middle point of SO is fixed.

And since the circum-⊙ is given, ∴ the radius of the nine-points ⊙ is given [Ex. 33 (ii), p. 302]. Hence the nine-points ⊙ is fully determined.

38. Take the figure of p. 243.

Let BC be the fixed base of the △ABC, having its vert. ∠ BAC constant in magnitude. Then the circum-⊙ is fixed in position and magnitude [Ex. 8, p. 299]. Hence the ⊙ about the pedal △DEF is fixed in magnitude, for its radius is half that of the circum-⊙ [Ex. 33, p. 302].

Now the ∠ FDE at the ○ᶜᵉ is constant, for it is the suppᵗ. of twice the vert. ∠ A [Ex. 20, p. 243, Cor.]. ∴ the chord FE is of constant length [III. 26, 29].

39. Take the fig. of p. 298.

Now the base BC is given, and the vert. ∠ BAC is constant, ∴ the circum-⊙ is fixed in magnitude and position.

But ABC is the pedal △ of the △$I_1I_2I_3$ [Ex. 2 (v), p. 298],

∴ the circum-⊙ of the △ABC is the nine-points ⊙ of $I_1I_2I_3$.

Hence if AI_1, and I_2I_3 cut the ⊙ about ABC at X and Y, these are the middle points respectively of II_1 and I_2I_3 [Ex. 32, p. 301].

But since XAY is a rt. ∠ [Ex. 2, p. 35], ∴ XY is a diam., and its middle point N is the centre of the ⊙ about ABC, and is a fixed point.

But X is a fixed point (the middle point of the arc BC, since ∠ BAX = ∠ CAX), ∴ Y is a fixed point.

Draw YS perp. to I_2I_3 meeting IN produced at S.

Then S is the centre of the ⊙ about $I_1I_2I_3$, for this centre must lie both in YS and IN produced [Ex. 33, p. 302].

And SN = NI : also YN = XN (proved), and ∠ SNY = vert. opp. ∠ XNI ;

$$\therefore\ SY = IX\ [\text{I. } 4] = \tfrac{1}{2} II_1 \text{ (proved above).}$$

But II_1 is a diam. of the ⊙ about the △BIC ; and this is a fixed ⊙, for the base and vert. ∠ are constant [Ex. 36, p. 246]. Hence SY is constant; and as Y has been shewn to be a fixed point, the locus of S is a ⊙ about Y as centre.

II. Miscellaneous Examples. **Page 303.**

1. Let ABCD be the quadrilateral. Let the bisectors of the $\angle^s$ A, B meet at X; of the $\angle^s$ B, C at Y; of the $\angle^s$ C, D at Z; and of the $\angle^s$ D, A at V.

Then the ext. $\angle$ AXY $= \frac{1}{2}A + \frac{1}{2}B$ [I. 32].

Also the ext. $\angle$ YZD $= \frac{1}{2}C + \frac{1}{2}D$.

$\therefore$ the $\angle^s$ AXY, YZD $= \frac{1}{2}A + \frac{1}{2}B + \frac{1}{2}C + \frac{1}{2}D =$ two rt. angles.

$\therefore$ the $\angle^s$ YXV, YZV $=$ two rt. angles;

$\therefore$ the points X, V, Z, Y are concyclic.

2. Let AB, BC, CD ... be the sides of the figure, O the point of intersection of the bisectors of the angles, and OX, OY, OZ, ... the perps. on AB, BC, CD ....

Then from the $\triangle^s$ OBX, OBY, we have OX $=$ OY [I. 26].

And from the $\triangle^s$ OCY, OCZ, we have OY $=$ OZ [I. 26].

And so on for each pair of sides. $\therefore$ the perps. OX, OY, OZ, ... are all equal. Hence a $\odot$ may be inscribed in the figure.

3. Circumscribe an equilat. triangle ABC about the given $\odot$.

Take O the centre, and join OA, OB, OC.

Inscribe a $\odot$ in each of the equal $\triangle^s$ AOB, BOC, COA.

These $\odot^s$ will touch one another and the original $\odot$.

4. Take the figure of p. 298.

Draw I_1D_1 and I_2E_2 perp. to BC, AC respectively, and let these lines meet at P.

Then from the $\triangle I_1D_1C$, the $\angle I_2I_1P = \frac{1}{2}C$.

And from the $\triangle I_2E_2C$, the $\angle I_1I_2P = \frac{1}{2}C$.

Hence $PI_1 = PI_2$, and P lies on the line which bisects I_1I_2 at rt. angles.

Similarly the intersections of I_2E_2, I_3F_3 and of I_3F_3, I_1D_1 lie on the lines which bisect I_2I_3 and I_3I_1 at rt. angles.

Hence I_1D_1, I_2E_2, I_3F_3 meet at the centre of the $\odot$ circumscribed about the $\triangle I_1I_2I_3$.

5. If the circum-⊙ and the vert. ∠ are given in magnitude, the length of the base is determined [III. 26, 29]. Hence the problem is reduced to that solved in Ex. 10, p. 299.

Or, otherwise. Describe the circum-⊙ with the given radius, and from any point P in its ○^{ce} draw the chords PB, PC each subtending at the centre an ∠ equal to the given ∠. Join BC; and on the side of BC remote from P, draw EF par^{l}. to BC at a distance from it equal to the radius of the in-⊙. From centre P with radius PB intersect EF at I, and produce PI to meet the ○^{ce} at A.

Then ABC shall be the required △. Join BI.

For the ∠ BAC = the given ∠ [III. 20]. Also the ∠ BAP = the ∠ CAP.

And since PB = PI, ∴ the ∠ PIB = the ∠ PBI.

But the ∠ PIB = the ∠s IAB, IBA; and ∠ IAB = ∠ PBC [III. 27].

Hence ∠ PBI = the ∠s IBA, PBC;

∴ ∠ IBA = ∠ IBC; that is, IB bisects the ∠ ABC.

Hence I, the intersection of the bisectors of the ∠s ABC, BAC, is the centre of the inscribed ⊙; and its distance from BC is equal to the given radius.

6. Let BC be the given base, and let BP of unlimited length make the given angle with BC; also let the st. line K be the given length of II_1 [See fig., p. 298].

Analysis. Since BI and BI_1 are the internal and external bisectors of the ∠ ABC, ∴ the ∠ IBI_1 is a rt. angle. And since II_1 is of given length, ∴ the middle point Q of II_1 lies on a ⊙ having B as centre and a radius equal to half of K [III. 31]. Also Q, the middle point of II_1, lies on the ○^{ce} of the circum-⊙ about the required △ [Ex. 15, p, 299].

∴ Q also lies on the st. line which bisects BC at rt. angles.

Hence the following construction.

From centre B and radius equal to $\frac{1}{2}$K describe a ⊙.

Bisect BC at rt. angles by a st. line which cuts this ⊙ (on the side of BC remote from BP) at Q.

Through B, Q, and C describe a ⊙: this will be the circum-⊙ of the required △, and will cut BP at the vertex A.

7. If the circum-⊙ of a △ is drawn, and the length of the base given, the magnitude of the vert. ∠ is determined [III. 28, 27].

Let P, Q be the given points. [In the figure taken P and Q are within the ⊙]. On PQ describe a segment of a ⊙ containing an angle equal to the vert. ∠ previously determined, and cutting the given ⊙ at A (or A′).

Join AP, AQ, and produce them to meet the ○^{ce} at B and C.

Then ABC is the required △.

There will be *two* solutions, *one* solution, or *no* solutions, according as the arc of the segment cuts the given ⊙, touches it, or falls wholly within it.

8. See figure, p. 298.

Since IBI_1, ICI_1 are rt. angles, the ⊙ about the △BIC passes through I_1, and II_1 is its diam., so that S_1 is the middle point of II_1. Similarly S_2, S_3 are the middle points of II_2, II_3.

Hence S_1S_2 is par^l. to I_1I_2 [Ex. 2, p. 104], and S_1S_2 is half of I_1I_2 [Ex. 3, p. 105].

Now by Ex. 4, p. 105, the △IS_1S_2 is one-fourth of the △II_1I_2; and similarly, the △^{s} IS_2S_3, IS_3S_1 are respectively one-fourth of the △^{s} II_2I_3, II_3I_1:

hence the △$S_1S_2S_3$ is one one-fourth of the △$I_1I_2I_3$.

9. Draw CD perp. to BC to meet the circum-⊙ at D. Join AD, BD.

Then BD is a diam., and AD is perp. to BA [III. 31].

But CO produced is also perp. to AB [Hyp.]; ∴ AD and CO are par^l. Similarly AO and DC are par^l.: ∴ AO = DC [I. 34].

But, since BCD is a rt. ∠, $BD^2 = BC^2 + CD^2$ [I. 47],

$$\text{or, } d^2 = BC^2 + AO^2.$$

10. Let ABCD ... be the regular polygon, O the centre of its inscribed ⊙, and P the given point within it. Let a denote the length of each side, r the radius of the inscribed ⊙, and $p_1, p_2, \ldots p_n$ the perps. drawn from P to the sides.

Then the area of the polygon = n-times the $\triangle$ AOB

$= n \cdot \frac{1}{2}ra$ [I. 41]

$= \frac{1}{2}nr \cdot a$ [II. 1]

= the area of a $\triangle$ on the base a having an alt. nr.

Similarly the area of the polygon

$= \triangle$ APB + $\triangle$ BPC + $\triangle$ CPD + ...

$= \frac{1}{2}p_1 a + \frac{1}{2}p_2 a + \frac{1}{2}p_3 a + \ldots \frac{1}{2}p_n a$ [I. 41]

$= \frac{1}{2}(p_1 + p_2 + \ldots + p_n)\,a$ [II. 1]

= the area of a $\triangle$ on the base a having an alt. $p_1 + p_2 + \ldots + p_n$.

But equal $\triangle^s$ on equal bases have equal altitudes,

$\therefore\ p_1 + p_2 + p_3 + \ldots + p_n = nr.$

11. Let ABCD ... be the regular polygon of n sides, of which O is the centre; and let PQ be the given st. line. Circumscribe a ⊙ about the polygon, and at the vertices A, B, C, D, ... draw tangents, thus forming another regular polygon of n sides, having the same centre O. Draw a tangent MN to the ⊙ par[l]. to PQ, and let T be its point of contact.

Then it may be shewn [I. 26] that the perps. from A, B, C, ... to MN are respectively equal to the perps. drawn from T to the corresponding sides of the outer polygon.

Hence the sum of the perps. from A, B, C, ... to MN = n times the perp. from O to MN [Ex. 10, p. 304].

$\therefore$ the sum of the perps. from A, B, C ... to the par[l]. line PQ = n times the perp. from O on PQ.

12. For the area of the quadrilat. is equal to the sum of the four $\triangle^s$ whose vertices are at the centre of the ⊙, and whose bases are the sides of the quadrilat. And if the lengths of these sides are given, their perp. distances from the centre are the same in all positions [III. 14].

But the areas of the four $\triangle^s$ depend only on the lengths of their bases and altitudes: hence the area of the quadrilat. is independent of the order in which the sides are placed.

13. Take the figure of p. 302.

Given O, N, and X, to draw the △ABC.

Join ON, and produce it to S, making NS equal to ON: then S is the centre of the circum-⊙ [Ex. 33, p. 302]. Join SX, and draw PXQ perp. to SX. Then the base BC must lie in PQ.

Through O draw DOR perp. to PQ: then the vertex A must lie in DR. Join XN, and produce it to meet DR in *a*: then X*a* is the diam. of the nine-point ⊙, and is therefore equal to the radius of the circum-⊙.

From centre S with radius equal to X*a* describe a ⊙ cutting PQ in B, C, and DR in A.

Then ABC is clearly the required triangle.

14. For suppose any two consecutive sides AB, BC of an inscribed polygon are unequal. Let P be the middle point of the arc AC.

Then AP, PC are together greater than AB, BC [Ex. 19, p. 264]; and △APC is greater than △ABC, for it has a greater altitude.

Hence there is an inscribed polygon which has a greater perimeter and a greater area than the given polygon.

∴ an inscribed polygon cannot have the maximum perimeter and maximum area unless every pair of consecutive sides are equal; that is, unless it is regular.

15. See fig., p. 273.

Suppose AG, AK two fixed tangents, touching a ⊙ at G and K.

Required to draw BC so that the sum of the lines BG, BC, CK may be a minimum.

Now since BH = BG, and CH = CK,

∴ the sum of BG, BC, CK = twice BC.

And it may be shewn that BC is a minimum when it touches the ⊙ at the middle point of the arc GK.

Hence arguing as in the last example we see that the points of contact of the sides of a circumscribed polygon of minimum perimeter must lie at equal intervals along the ○^{ce} of the inscribed ⊙. That is, the polygon must be regular. And since the polygon may be divided into △^{s} having a common vertex at the centre, and since the altitudes of these △^{s} are all equal to the radius of the in-⊙, ∴ the area of the polygon is a minimum when the perimeter is a minimum.

16. Let MAN be the given vert. ∠. Along AM, AN take AP, AQ each equal to half the given sum of the sides containing the vert. ∠.

Let ABC be one △ of the system. Then clearly PB = CQ.

At P and Q draw PX, QX perp. to AM and AN. Then X is a fixed point; and it may be shewn [I. 47 or Ex. 12, p. 99] that PX = QX. Hence the △^{s} BPX, CQX are identically equal [I. 4].

∴ the ∠ PXB = the ∠ QXC: to each add the ∠ PXC;

then the ∠ BXC = the ∠ PXQ.

But since the ∠^{s} at P and Q are rt. angles, the ∠^{s} PXQ, PAQ are supplementary; ∴ the ∠^{s} BXC, BAC are supplementary.

∴ X is on the ⊙ about the △ABC. Thus the ⊙ passes through two fixed points A and X. Hence the locus of the centre is the st. line bisecting AX at rt. angles.

17. In an equilat. △ the centroid, orthocentre, and centre of the circum-⊙ are at the same point O. Let AO produced meet BC at D and the ○^{ce} at E. Join PE, PD, PO.

Then OE = OA = twice OD [Ex. 4, p. 113]: ∴ OD = DE.

Let r denote the radius of the circum-⊙.

Then from △APE, $PA^2 + PE^2 = 2OA^2 + 2OP^2$ [Ex. 24, p. 161]

$$= 4r^2.$$

And from △BPC, $PB^2 + PC^2 = 2BD^2 + 2PD^2$.

By addition, $PA^2 + PB^2 + PC^2 + PE^2 = 4r^2 + 2BD^2 + 2PD^2$.

Again from △OPE, $r^2 + PE^2 = 2OD^2 + 2PD^2$.

By subtraction, $PA^2 + PB^2 + PC^2 - r^2 = 4r^2 + 2BD^2 - 2OD^2$

To each of these equals add r^2, or $4OD^2$.

Then $$PA^2 + PB^2 + PC^2 = 4r^2 + 2BD^2 + 2OD^2$$

$$= 4r^2 + 2r^2 \text{ [I. 47]}$$

$$= 6r^2.$$

BOOK VI.

Page 329.

1. Let AC, BD the diag^{ls}. of a quad^{l}. intersect in E.

Then $\triangle$ABE : $\triangle$BEC = AE : CE

= $\triangle$ADE : $\triangle$DEC [VI. 1].

2. Let the three par^{l}. st. lines cut one st. line in A, B, C and another in D, E, F. If ABC is par^{l}. to DEF,

AB = DE and BC = EF;

∴ AB : BC = DE : EF.

If not, through A draw AGH par^{l}. to DEF cutting BE and CF in G and H. Then AG = DE, and GH = EF.

Also AB : BC = AG : GH [VI. 2];

∴ AB : BC = DE : EF.

3. Because EF, EG are par^{l}. to bases AC, AD;

∴ AE : EB = CF : FB [VI. 2],

and AE : EB = DG : GB;

∴ CF : FB = DG : GB.

∴ FG is par^{l}. to base CD.

4. Draw CK par^{l}. to DF cutting AB in K.

Then BD : DC = BF : FK [VI. 2].

But ∠AFE = ∠AEF; ∴ AF = AE.

And AF : FK = AE : EC [VI. 2];

∴ FK = EC.

Hence BD : DC = BF : CE.

5. Let AD produced cut BC in G. Then $\triangle$^{s} ADB, GDB are identically equal [I. 26]. ∴ AD = DG; ∴ the par^{l}. through D to BGC bisects AC [VI. 2].

6. Let BE, CF cut the median AD in K.

Then AF : FB = △AKF : △FKB [vi. 1],

and AF : FB = △ACF : △FCB;

∴ AF : FB = △ACK : △BKC [Cf. v. 15].

Similarly AE : EC = △ABK : △BKC.

But because BD = DC, ∴ △BKD = △CKD, and △BAD = △CAD,

∴ △ABK = △ACK.

Hence AF : FB = AE : EC;

∴ EF is par[l]. to BC.

7. Draw PQ par[l]. to BC cutting AC in Q. Produce QC to R, making CR = QC. Join PR cutting BC in X.

Then, because CX is par[l]. to PQ,

∴ PX : XR = QC : CR;

but QC = CR, ∴ PX = XR.

8. Let G be the required pt., so that △BGC = △CGA = △AGB. Let AG, BG, CG produced cut the sides in D, E, F.

Then AF : FB = △AGF : △BGF

= △ACF : △BCF

= △AGC : △BGC;

∴ AF = FB. Similarly AE = EC, and BD = CD. Hence G is the centroid of ABC [Ex. 4, p. 113].

Page 331.

1. Because DE bisects ∠ADB;

∴ BE : EA = BD : DA.

Because DF bisects ∠ADC,

∴ CF : FA = CD : DA.

But BD = CD. ∴ BE : EA = CF : FA.

∴ EF is par[l]. to BC [vi. 2].

2. Let AB be the given st. line. On AB describe a triangle ABC, having BC double of AC [i. 22].

Let CD bisect the ∠ ACB, and meet AB in D.

Then $\qquad$ BD : DA = BC : CA ;

$\therefore$ BD = twice DA.

Bisect BD in E. $\quad$ Then AD = DE = EB.

3. Let AD bisect ∠ BAC. Take I in AD, so that

AI : DI = BA + AC : BC.

Because AD bisects ∠ BAC,

$\therefore$ BD : DC = BA : AC,

$\therefore$, *componendo*, $\qquad$ BD : BC = BA : BA + AC.

$\therefore$, *alternately*, $\qquad$ BD : BA = BC : BA + AC = DI : AI.

$\therefore$ BI bisects ∠ ABC. Similarly CI bisects ∠ ACB. $\therefore$ I is the centre of the inscribed ⊙.

4. Let the bisectors of ∠ˢ A and C meet at X in BD.

Then $\qquad$ DA : BA = DX : BX = DC : BC.

$\therefore$, *alternately*, $\qquad$ DA : DC = BA : BC.

Let Y divide AC in this last ratio. Then DY, BY are the bisectors of ∠ˢ D and B, and therefore meet in AC.

5. Let AB be the given base, and let the st. lines X, Y be in the given ratio. From A draw AP = X, and produce it to Q so that PQ = Y. Join QB, and draw PR parˡ. to QB, cutting AB in R.

Then AR : RB = AP : PQ = given ratio.

$\therefore$ R is the pt. where the bisector of the vert. angle is to cut AB. Proceed then as in Ex. 3, p. 223.

6. Let BI, CI, the bisectors of B and C, intersect in I. Join AI. And produce AI to cut BC in D.

Then $\qquad$ BA : BD = AI : ID $\quad$ [VI. 3].

Also $\qquad$ CA : CD = AI : ID.

$\therefore$ BA : BD = CA : CD,

or, $\qquad$ BA : CA = BD : CD.

$\therefore$ AD bisects the ∠ A.

7. Because arc AC = arc AD, ∴ ∠ CFE = ∠ DFE; that is, FE bisects ∠ CFD. Similarly GE bisects ∠ CGD.

∴ CG : DG = CE : DE = CF : DF.

∴ CG : CF = DG : DF.

Page 333.

1. Join BP. Then ∠ APB is a rt. ∠ [III. 31]. And PA bisects the ∠ CPD, ∴ PB bisects the adj. supplementary angle.

∴ CP : PD = CA : AD [VI. 3],

and CP : PD = CB : BD [VI. A];

∴ CA : AD = CB : BD; or, AC : CB = AD : DB.

2. Because AB = AE, ∴ ∠ CBA = ∠ DEA. And ∠ BAC = ∠ EAD. ∴ △ˢ ABC, AED are identically equal [I. 26]. Since AC bisects ∠ BAD, and AE is at rt. ∠ˢ to AC; ∴ AE bisects the adj. supplementary angle.

∴ BC : CD = BA : AD [VI. 3]; and BE : ED = BA : AD [VI. A].

∴ BE : ED = BC : CD.

But DE = BC. ∴ DE or BC is a mean proportional between BE and CD.

3. Let BI_1, CI_1, the bisectors of the ext. ∠ˢ B and C, intersect in I_1. Join AI_1 cutting BC in D.

Then $AB : BD = AI_1 : I_1D$ [VI. A].

Also $AC : CD = AI_1 : I_1D$ [VI. A].

∴ AB : BD = AC : CD; or, AB : AC = BD : DC.

∴ AD bisects int. ∠ A [VI. 3].

Page 335.

1. Let ABCD be a trapezium, in which AD is par[l]. to and double of BC. Let AC, BD cut at E. Then △ˢ AED, CEB are equiangular to one another.

∴ AE : EC = DE : EB = AD : CB.

∴ AE = twice EC, and DE = twice EB.

2. Since DE = EA, and BG = GA,

∴ GE is par^l. to BD [VI. 2]; and △^{s} AGE, ABD are equiangular

∴ AG : AB = GE : BD;

but AG is half of AB, ∴ GE is half of BD.

Similarly HF is half of BD; ∴ GE = HF.

3. Because △^{s} ABF, CEF are equiangular,

∴ EF : BF = EC : BA.

And because △^{s} ABG, EDG are equiangular,

∴ EG : AG = ED : AB.

But CE = ED.

∴ EF : BF = EG : AG.

∴ GF is par^l. to AB [VI. 2].

4. The △^{s} BAE, AED are identically equal [I. 4].

∴ ∠ AEF = ∠ ADE.

And ∠ EAF is common to △^{s} AFE, AED, ∴ △^{s} AFE, AED are equiangular [I. 32].

∴ AF : AE = AE : AD.

5. Because △^{s} AHK, ADC are equiangular:

∴ AH : AD = KH : CD:

that is, EK : EF = KH : GH.

∴ EH is par^l. to base GF of △GKF. [VI. 2].

Again, let GE meet CA in M. Then, because △^{s} MEK, MGC are equiangular,

∴ EK : GC = KM : CM.

Let FH meet CA in N. Then HK : FC = KN : CN.

But EK : GC = HK : FC;

∴ KM : CM = KN : CN.

CK : KM = CK : KN [Cf. v. 13];

∴ KM = KN.

That is, M and N *coincide*.

6. Because EF is par^l. to DA, ∴ ∠ FEB = ∠ DAE. But

∠ DAE = ∠ FCE, in the same segment.

∴ ∠ FEB = ∠ FCE. And ∠ at F is common to the △^{s} FEB, FCE, which are therefore equiangular [I. 32].

∴ FB : FE = FE : FC.

Page 339.

1. Let BC, AD be the par^l. sides of a trapezium ABCD, and let AC, BD cut in E. Then △^{s} EAD, ECB are equiangular.

∴ AE : EC = DE : EB [VI. 4].

2. Let PA, QB, RC cut in O. Then △^{s} OBA, OQP are equiangular.

∴ AB : PQ = OB : OQ [VI. 4].

Similarly, BC : QR = OB : OQ;

∴ AB : PQ = BC : QR.

3. Join PQ, PR. Then because OP is tangent at P,

∴ ∠ OPQ = ∠ ORP [III. 32].

And the angle at O is common to △^{s} OPQ, ORP. ∴ these △^{s} are equiangular. [I. 32].

∴ OR : OP = OP : OQ [VI. 4].

4. See fig. of I. 38. Let a par^l. to BF cut AB, AC, and DE, DF in X, Y, P, Q.

Then XY : BC = AY : AC,

and PQ : EF = DP : DE [VI. 4].

But AY : AC = DP : DE [Ex. 2, p. 329].

∴ XY : BC = PQ : EF.

But BC = EF. ∴ XY = PQ. ∴ △AXY = △DPQ [I. 38].

5. Because, in △^{s} POX, YOQ, ∠^{s} POX, YOQ are equal,

and PO : OX = YO : OQ;

∴ △^{s} POX, YOQ are similar [VI. 6]. ∴ ∠ OPX = ∠ OYQ.

∴ P, X, Q, Y are concyclic. [Converse of III. 21.]

6. Because △^{s} ACB, ADB are equal, ∴ DC, AB are par^l. [I. 39];

∴ DO : DB = CO : CA [VI. 2].

But, because AD is par^l. to OE,

∴ DO : DB = AE : AB [VI. 2].

Similarly CO : CA = BF : BA,

∴ AE : AB = BF : AB;

∴ AE = BF.

7. Because $\angle$s ABD, ACD are rt. $\angle$s, $\therefore$ AD is the diameter of the circumcircle of ABC [III. 31]. $\therefore \angle$ ADC $= \angle$ ABC.

But $\angle$ ADC $= \angle$ ACE, since each is the comp^t. of $\angle$ CAD [I. 32]. $\therefore \angle$ ACE $= \angle$ ABC. $\therefore \triangle$s ACE, ABC are equiangular [I. 32].

8. Because EF, AC are par^l.,

$\therefore$ CF : FD = AE : ED [VI. 2].

But $\triangle$s AEC, DEB are equiangular;

$\therefore$ AE : ED = AC : BD [VI. 4];

$\therefore$ CF : FD = AC : BD,

and, *alternately*, CF : AC = FD : BD;

also $\angle$ ACF = $\angle$ BDF, $\therefore \triangle$s ACF, BDF are similar [VI. 6];

$\therefore \angle$ AFC = $\angle$ BFD.

9. Because $\triangle$s RPQ, RAB are equiangular,

$\therefore$ PQ : AB = PR : AR [VI. 4].

And because $\triangle$s SPQ, SDC are equiangular,

$\therefore$ PQ : CD = PS : DS.

But AB = CD; $\therefore$ PR : AR = PS : DS.

$\therefore$ SR is par^l. to AD [VI. 2].

10. Draw EG par^l. to AB, cutting BC in G

Then AB : AC = EG : EC = EG : DB = EF : DF.

11. Let X, Y, Z be the st. lines, whose ratios are to be equal to those of the perps. on BC, CA, AB. Draw CD perp. to BC, on the same side as A and equal to X. Draw CE perp. to CA, on the same side as B and equal to Y. Draw DQ, EQ par^l. to BC, AC to meet in Q.

Then, by similar $\triangle$s, the perps. from any point in CQ, or CQ produced, upon BC, CA are in the ratio CD : CE; that is, X : Y. Similarly a line can be drawn, such that the perp. from any pt. in it upon BC, AB are in the required ratio X : Z.

Hence the pt. of intersection of the two lines so drawn is the pt. required.

Page 344.

1. In the figure of Prop. 8, because the $\triangle^s$ BCA, BAD are similar,

$$\therefore \text{BC} : \text{CA} = \text{BA} : \text{AD}.$$

2. The tangent intercepted between two par^l. tangents subtends a rt. $\angle$ at the centre. [Ex. 10, p. 197.] And the perp. from the centre upon it is the radius.

Page 345.

1. Let AB be the st. line. Draw a st. line ACD; cut off AC = half AB, and CD = twice AB. Join DB, and draw CF parallel to DB. AF shall be a fifth of AB.

2. Let AB be the st. line. Draw a st. line ACDE. Cut off CD = double of AC and DE = half of AC. Join EB: and draw CF par^l. to EB. CF shall be two-sevenths of AB.

Page 346.

Let AB be the given line, and ACD any other line, divided at C so that AC : CD = the given ratio. Join DB and draw CH par^l. to DB. This must cut AB in some pt. H, so that AB is divided *internally* at H in the required ratio. Next, on CA (produced, if necessary) take CD′ = CD. Join D′B, and draw CH′ par^l. to D′B. This must cut AB produced in some pt. H′, *unless* D′ *coincides with* A. Thus AB is divided *externally* at H′ in the required ratio, unless that ratio is a ratio of equality. [Or see p. 384.]

Page 347.

1. Join BC. Then because rt. $\angle$ ACB = rt. $\angle$ ABD, and $\angle$ A is common to $\triangle^s$ ACB, ABD, $\therefore$ the $\triangle^s$ are equiangular [I. 32].

$$\therefore \text{AD} : \text{AB} = \text{AB} : \text{AC} \quad [\text{VI. } 4].$$

2. Because $\angle$ BCA = twice $\angle$ BAC, $\therefore$ the $\angle$ BCD = $\angle$ BAC. And $\angle$ B is common to $\triangle^s$ BCD, BAC. $\therefore$ the $\triangle^s$ are equiangular;

$$\therefore \text{BA} : \text{BC} = \text{BC} : \text{BD} \quad [\text{VI. } 4].$$

3. Because AD touches ACB at A, $\therefore$ $\angle$ BAD = $\angle$ ACB [III. 32]; and because AC touches ADB at A, $\therefore$ $\angle$ BAC = $\angle$ BDA.
$\therefore$ $\triangle^s$ BCA, BAD are equiangular [I. 32];

$$\therefore \text{BC} : \text{BA} = \text{BA} : \text{BD} \quad [\text{VI. } 4].$$

Page 348.

1. The △s ADE, CFD are similar,

∴ AE : AD = CD : CF = AB : CF.

2. Because ∠s ABD, AEC are in same segment, they are equal. And ∠ BAD = ∠ EAC. ∴ △s ABD, AEC are equiangular.

∴ BA : AD = EA : AC [VI. 4].

3. Join QR, PC. Then PC bisects QR at rt. angles [Ex. 2, p. 196]. ∴ ∠ QRT = compt. of ∠ PCR = ∠ CPR. And ∠s PRC, RTQ are rt. angles. ∴ △s PRC, RTQ are equiangular [I. 32].

∴ PR : RC = RT : TQ [VI. 4].

Pages 349, 350.

1. Let P be in the side CD, Q in the side BC. Then by similar △s EPD, EAB,

EP : EA = ED : EB.

And, by similar △s EAD, EQB,

EA : EQ = ED : EB.

∴ EP : EA = EA : EQ.

2. On AC describe a semicircle AQC. At B, make ∠ ABQ = $\frac{1}{2}$ a rt. ∠. Draw QP perp. to AC. Then PB = PQ; and since PQ is perp. to AC, it is a mean proportional between PA and PC.

3. From the similar △s ODC, CDE, OD : DC = DC : DE;

that is, AO : DC = DC : DE.

4. Let O be the centre. Then △ OBC is equilateral. Produce OC to D, making CD = OC. Join BD. Then ∠s AOB, BCD, being the supplements of equal ∠s, are equal. Hence △s AOB, BCD are identically equal [I. 4]. ∴ ∠ ABO = ∠ BDC. And ∠ at A is common to △s OAB, BAD: ∴ these △s are equiangular [I. 32];

∴ OA : AB = AB : AD [VI. 4];

that is, AB is mean proportional between BC and BC + CA

5. In the rt. angled $\triangle^s$ DEB, GEA, the comp^{ts}. of equal $\angle^s$ at E are equal; ∴ the $\angle$ DBE = the $\angle$ DGC; ∴ $\triangle^s$ DEB, DCG are similar.

∴ DE : DB = DC : DG.

But DB : DF = DF : DC [VI. 8].

∴, *ex æquali*, DE : DF = DF : DG. That is, DG is a third proportional to DE and DF.

6. Let P be the pt. of contact of the circles; then APB is a rt. $\angle$ [Ex. 21, p. 237]. Produce AP, BP to meet the ○^{ces} in C and D; then AD, BC are diameters, since the $\angle^s$ APD, BPC are rt. $\angle^s$. It is easily seen that $\triangle^s$ DAB, ABC are equiangular,

hence DA : AB = AB : BC.

7. Let C, D be two given pts. on AB. On AD, BC as diameters describe semicircles cutting in P. Draw PX perp. to AB.

Then XA : XP = XP : XD [VI. 8].

And XP : XB = XC : XP.

∴ *ex æquali*, XA : XB = XC : XD.

∴ X is the required pt

8. Because AB : AC = AC : AD;

∴ AB : AB − AC = AC : AC − AD;

∴ AB : AC = BC : CD.

Again AB : AE = AE : AD.

∴ in $\triangle^s$ ABE, AED, the sides about the common $\angle$ at A are proportional. ∴ these $\triangle^s$ are similar [VI. 6].

∴ BE : ED = AB : AE = AB : AC = BC : CD.

∴ CE bisects $\angle$ BED.

9. Let ABC be a $\triangle$. Produce BC to F, making CF a third proportional to BC, CA. Join FA, and draw CE par^l. to FA cutting AB in E. Draw ED par^l. to AC cutting BC in D.

Then BD : DE = BC : CA = CA : CF = DE : DC.

∴ DE is a mean proportional to BD, DC, and is par^l. to AC.

10. Because OF = OA,

∴ OE : OF = OE : OA

= OA : OD, from similar △s EOA, AOD

= OD : OG, from similar △s AOD, DOG

11. The ∠s ABF, BEA are subtended by equal chords AB, AC of ⊙ ABEC; ∴ they are equal. And the ∠ at A is common to the two △s ABF, AEB. ∴ these △s are equiangular [I. 32]; ∴ AE : AB = AB : AF. ∴ AB, or AD, is a mean proportional between AE, AF.

Page 353.

1, 2. Let the parms. ABCD, EFGH; or the △s BCD, FGH be equal in area, and let BC : FG = GH : CD.

Then, if ∠ FGH is greater than ∠ BCD, make ∠ FGK = ∠ BCD, and GK = GH.

Then BC : FG = GK : CD, ∴ △ BCD = △ FGK [VI. 15].
But △ BCD = △ FGH. ∴ △ FGK = △ FGH. ∴ K is on EH [I. 39].
∴ ∠ FGK = ∠ GKH = ∠ GHK [I. 5] = suppt. of ∠ FGH. Hence the ∠s BCD, FGH are either equal or supplementary. In either case, the parms. ABCD, EFGH have their angles respectively equal.

3. Because AC, BD meet the parls. AB, CD, ∴ the △s OAB, OCD, are equiangular to one another.

∴ AO : OC = BO : OD.

∴ the sides of the △s AOD, BOC about their equal ∠s AOD, BOC are reciprocally proportional. ∴ △ AOD = △ BOC.

4. The △s CAE, CDB are similar, because AE, BD are parl.

∴ CA : CD = CE : CB.

∴ △ ABC = △ CDE.

5. The △s ADE, AFG are similar,

hence AD : AE = AF : AG.

Again, ∠ EAD = ∠ GAF, to each add ∠ EAG; then

∠ DAG = ∠ EAF ;

and the sides about the equal ∠s DAG, EAF are reciprocally proportional.

∴ △ DAG = △ EAF.

6. Because GE is par^l. to AC,

∴ AG : AD = CE : CD = FA : BA. Hence the sides about the common ∠ A of the △^{s} DAF, GAB are reciprocally proportional.

7. Let BAC be given △. Produce BA, CA to D and E, so that AD = AE = mean proportional between BA, AC [VI. 8].

Then BA : AD = AE : AC.

∴ △ DAE = △ ABC, and DAE is isosceles, with vert. angle = ∠ A.

8. AB : AC = AC : AD [VI. 8];

that is, AZ : AC = AY : AD.

∴ sides about the equal ∠^{s} ZAD, CAY of the △^{s} ZAD, CAY are reciprocally proportional. ∴ △ ZAD = △ CAY. Similarly △ ZBD = △ CBX. ∴ △ ABZ = △ CAY + △ BCX.

Page 357.

1. Let chords AB, CD intersect in O. Join AC, BD. Then ∠^{s} OAC, ODB in same segment are equal, and ∠^{s} OCA, OBD in same segment are equal. ∴ △^{s} OAC, ODB are equiangular [I. 32].

∴ OA : OC = OD : OB [VI. 4].

∴ rect. OA, OB = rect. OC, OD.

2. Let ABC be right ∠^{d} at A, and AD perp. on BC. Then, because △^{s} ABC, DAC are similar [VI. 8],

∴ AB : BC = DA : AC.

∴ rect. AB, AC = rect. BC, DA.

3. Let ABCD be the given rect., and EF the given line. To EF, AB, BC find a fourth proportional. Draw EG perp. to EF and equal to this fourth proportional.

Then EF : AB = BC : EG.

∴ rect. EF, EG = given rect. AB, BC.

4. From the similar △^{s} FAE, FCB,

FE : FB = FA : FC.

And from the similar △^{s} FAB, FCG,

FA : FC = FB : FG.

∴ FE : FB = FB : FG.

∴ rect. FE, FG = sq. on FB.

5. Let ABC be given △; DE the given st. line. Draw AN perp. to BC. Bisect DE in F. Draw FG perp. to DE and equal to the fourth proportional to DE, BC, AN. Then

rect. DE, FG = rect. BC, AN. ∴ △GDE = △ABC [I. 41].

6. Because ∠ˢ ACB, ABD are rt. ∠ˢ, and ∠A is common to △ˢ ABC, ADB. ∴ these △ˢ are equiangular,

∴ AC : AB = AB : AD [VI. 4];

∴ rect. AC, AD = sq. on AB = constant.

7. Because AD bisects the ext. ∠ at A, ∴ ∠BAE = ∠CAD; and the ∠ACD = the ∠BEA.

∴ the △ˢ AEB, ACD are equiangular [I. 32];

∴ AE : AB = AC : AD.

∴ rect. AE, AD = rect. AB, AC.

8. Join B to F, the other extremity of the diam.

Then rect. AB, AD = sq. on diameter [Ex. 6, p. 357]

= rect. AC, AE.

∴ AB : AC = AE : AD [VI. 16].

∴ △ˢ ABC, AED are similar [VI. 6].

9. Let C be the centre, and ACB the diameter. Then, because CQ, CR bisect adj. supplementary ∠ˢ ACP, BCP, ∴ ∠QCR is a rt. ∠. And CP is perp. to QR.

∴ QP : CP = CP : PR [VI. 8].

∴ rect. QP, PR = sq. on CP = constant.

10. The ∠AEB = ∠ACB in same segment = ∠ABD.

∴ △ˢ AEB, ABD are equiangular;

∴ AE : AB = AB : AD [VI. 4],

∴ rect. AE, AD = sq. on AB.

11. Let the tangent at A meet BC in D. Then, because the tangents DA, DB, DC are equal, the ⊙ with centre D and radius DA passes through B and C. But SA, being the line of centres of the given ⊙ˢ is perp. to DA. Hence SA is the tangent to the ⊙ circumscribing ABC. ∴ sq. on tangent SA = rect. of the segments of the secant SB, SC.

12. Let AD be the median, and AL the perp. on base BC. From BC cut off BX equal to the mean proportional between BD and BL. The perp. XY will bisect the $\triangle$ABC.

For BD : BX = BX : BL = BY : BA.

That is, the $\triangle$s ABD, XBY have their sides about the common $\angle$ B reciprocally proportional;

$\therefore$ $\triangle$XBY = $\triangle$ABD = half $\triangle$ABC.

Page 369.

Let ABCD, EFGH be two parms. Draw BM, CN perp. to AD and FQ, GR perp. to EH. Then parm. BMNC = parm. ABCD, and parm. QFGR = parm. EFGH.

But parm. BN : parm. FR in the ratio compounded of BC to FG and BM to FQ, for these parms. are equiangular.

$\therefore$ parms. BD, FH have to one another the ratio compounded of the ratios of their bases and of their altitudes. But $\triangle$ ABC = half parm. BD and $\triangle$EFG = half parm. FH. Hence the same is true of the $\triangle$s ABC, EFG.

Page 373.

1. The $\triangle$s AEB, EDA are identically equal; $\therefore$ $\angle$ AEB = $\angle$ EDA. $\therefore$ $\triangle$s AEO, ADE are equiangular [I. 32.];

$\therefore$ AD : AE = AE : AO [VI. 4.].

$\therefore$ rect. AD, AO = sq. on AE.

But it may be shewn, as in Ex. 6, p. **293**, that OD = ED = AE;

$\therefore$ rect. AD, AO = sq. on OD.

2. See fig. p. 282. Let AB be divided at C in extreme and mean ratio, so that rect. BA, BC = sq. on AC. Then AC = BD, and BD is the side of the regular decagon inscribed in the $\odot$ BDE [Ex. 6 (i), p. 283].

Page 375.

1. See fig. p. 374. Then BD : CD = fig. R : fig. Q; that is, in duplicate ratio of AB to AC [VI. 20].

2. The $\triangle$s ABC, DBA, DAC are similar and similarly described on BC, BA, AC. Hence fig. P : fig. Q = $\triangle$ABC : $\triangle$DAC, since each ratio is the duplicate ratio of BC to AC. ∴ if P = △ABC, then Q = △DAC; similarly it may be shewn that R = △DBA.

3. Since △s AGB, XGY are similar and similarly described on AG, XG; ∴ △AGB is to △XGY in the duplicate ratio of AG to XG; that is as sq. on AG is to sq. on XG. But AG = twice XG. ∴ △AGB = 4 times △XGY.

4. Let ABC, A′B′C′ be similar △s.

(i) Let AD, A′D′ be corresponding medians.

Then AB : BD = A′B′ : B′D′; and ∠B = ∠B′.

∴ △s ABD, A′B′D′ are similar, and AD homologous to A′D′.

∴ △ABC : △A′B′C′ = △ABD : △A′B′D′
= dupl. ratio of AD : A′D′.

(ii) Let IX, I′X′ be corresponding in-radii perp. to BC, B′C′.

Then IX : I′X′ = IB : I′B′, from similar △s IBX, I′B′X′;
= BC : B′C′, from similar △s IBC, I′B′C′:

But △ABC : △A′B′C′ = dupl. ratio of BC : B′C′,
= dupl. ratio of IX : I′X′.

(iii) Let S, S′ be the respective circumcentres.

Then SB : S′B′ = BC : B′C′; from similar △s SBC, S′B′C′.

∴ △s ABC, A′B′C′ are to one another in the dupl. ratio of the circum-radii.

5. The △s DBF, ABC are similar, DB and AB being homologous sides. [Ex. 20, Cor. ii. p. 243.]

∴ △ABC : △DBF = dupl. ratio of AB to DB
= $AB^2 : DB^2$.

∴ △ABC − △DBF : △DBF = $AB^2 - DB^2 : DB^2$ [v. 13];

or, quad[l]. AFDC : △DBF = $DA^2 : DB^2$ [I. 47].

6. [The question assumes that AB is greater than AC.]

From AB cut off AX a third proportional to BA, AC; and join CX.

Then BA : AX = dupl. ratio of BA to AC [Def.].

And BD : DC = dupl. ratio of BA to AC [Hyp.],

∴ BA : AX = BD : DC.

∴ CX is par[l]. to AD [VI. 2].

Again BA : AC = CA : AX;

Hence the △[s] BAC, CAX are similar [VI. 6].

∴ ∠ ABC = ∠ ACX = alt. ∠ CAD;

and the △[s] ABD, CAD have the ∠ D common; hence they are equiangular [I. 32];

∴ BD : DA = DA : DC [VI. 4].

7. Let ABC be the △. Draw the median AD; and from BC cut off BE a mean proportional between BD and BC. Draw EF par[l]. to CA. Then EF shall bisect the △ ABC.

For BD : BE = BE : BC = BF : BA.

∴ △ EBF = △ ABD [VI. 15]

= half △ ABC.

8. Let ABC be the △. Produce BC to D making BD double of BC, and from BD cut off BE a mean proportional between BD and BC. Join AD, and draw EF par[l]. to AC to meet BA produced at F.

Then BFE is the △ required. [Proof as in Ex. 7.]

9. Let AD, BE, CF meet in O.

Then BD : DC = △ BOA : △ AOC.

And BF : FA = △ COB : △ AOC,

and AE : EC = △ BOA : △ COB.

But △ BOA has to △ AOC the ratio compounded of the ratios of △ BOA to △ COB and of △ COB to △ AOC.

∴ BD has to DC the ratio compounded of the ratios of AE : EC and of BF : FA.

10. Let ABC be an isosceles △. From AB cut off AD equal to the mean proportional between AB or AC and BC. Draw DE par[l]. to BC. Then AB : AD = AD : BC;

also △ ABC : △ ADE in duplicate ratio of AB to AD;

that is, △ ABC : △ ADE = AB : BC.

11. Let P be the given pt., AB, AC the given st. lines. Describe an isosceles triangle HAK, having BAC as vertical ∠, and equal to the given rectilineal figure [vi. 25]. Draw PM, PN par[l]. to AC, AB cutting AB, AC in M and N respectively. Divide AH at X, so that rect. AX, XH = rect. AM, AN [Ex. 20, p. 266], and draw HC par[l]. to XN. The st. line CPB shall be the base of the required △.

From AK, cut off AY = XH; then YK = AX. Now, by parallels,

AM : MB = CP : PB = NC : AN

= XH : AX = AY : YK.

∴ BK is par[l]. to MY. Again rect. AX, AY = rect. AM, AN. ∴ MY is par[l]. to XN. But XN is par[l]. to HC. ∴ BK is par[l]. to HC. ∴ △ABC = △AHK = given rectilineal figure.

Page 380.

Let AD meet BC produced; and DA produced cut the circumcircle of ABC in E. Then ∠ DAC = ∠ EAB; hence ∠ BAD = ∠ EAC; and ∠ ABD = ∠ AEC [iii. 21]; ∴ △[s] ABD, AEC are equiangular;

∴ BA : AD = EA : AC.

∴ rect. BA, AC = rect. EA, AD.

. rect. BA, AC + sq. on AD = rect. EA, AD + sq. on AD

= rect. ED, AD [ii. 3].

= rect. BD, DC [iii. 36].

Page 383.

1. Draw AD perp. to base BC. Then rect. BA, AX = rect. contained by AD and diameter of circumcircle of BAX [Prop. C]. And rect. CA, AX = rect. contained by AD and diam. of ⊙ CAX. But BA = CA. ∴ diam. of ⊙ BAX = diam. of ⊙ CAX.

2. The △[s] ABD, ACD are identically equal [Ex. 12, p. 99]. Also A, B, D, C are concyclic [Converse of iii. 22].

∴ rect. BC, AD = rect. AB, CD + rect. AC, BD

= twice rect. AB, BD.

3. Let diagonals AC, BD intersect at rt. ∠[s] in E.

Then sum of rect[s]. of opp. sides = rect. AC, BD

= sum of rect[s]. AE, BE; BE, CE; CE, DE; DE, AE [ii. 1]

= twice sum of △[s] ABE, BCE, CDE, DAE

= twice area of ABCD.

4. Let BD bisect AC in E. Draw AX, CY perp. to BD.

Then rect. AB, AD = rect. contained by AX and the diam. of ⊙ [Prop. C].

Also rect. BC, CD = rect. contained by CY and the diam. of ⊙.

But AX = CY, from the identically equal △s EAX, ECY;

∴ rect. AB, AD = rect. BC, CD.

5. Draw AD perp. to BC and let X be any pt. in BC.

Then rect. AB, AX = rect. contained by AD and diam. of ⊙ about ABX.

Hence AD : AX = AB : diam. of ⊙ ABX [VI. 16].

Similarly AD : AX = AC : diam. of ⊙ ACX.

∴ AB : diam. of ⊙ ABX = AC : diam. of ⊙ ACX;

or, AB : AC = diam. of ⊙ ABX : diam. of ⊙ ACX.

6. Let BC be the given base. On BC describe a segment of a ⊙ containing an ∠ equal to given ∠. Let X, Y be the sides of given rectangle. To the diameter, X and Y, find a fourth proportional DA. Place DA in segment perp. to BC. Then BAC is the required △. [Prop. C.]

7. Let ABC, DEF be the two equal △s, and let AM, DN be perps. from the vertices A, D upon the bases BC, EF. Let PQ be the diameter of the ⊙ circumscribing the △s ABC, DEF.

Then rect. BA, AC = rect. PQ, AM.

And rect. ED, DF = rect. PQ, DN.

∴ rect. BA, AC : rect. ED, DF = AM : DN.

But △BAC = $\frac{1}{2}$BC, AM; and △DEF = $\frac{1}{2}$EF, DN.

∴ AM : DN = EF : BC.

∴ rect. BA, AC : rect. ED, DF = EF : BC.

8. Let P be on the arc BC of the circumcircle of the equilat. △ABC.

Then rect. PB, CA + rect. PC, AB = rect. PA, BC. [Prop. D.]

But BC = CA = AB. ∴ rect. (PB + PC), AB = rect. PA, AB.

∴ PB + PC = PA.

9. Because ∠ABD = ∠CBD; ∴ arc AD = arc CD; ∴ chord AD = chord DC. And because A, C are fixed,

∴ D is a fixed point, and AD is constant:

but rect. AB, CD + rect. BC, AD = rect. AC, BD,

or, rect. (AB + BC), AD = rect. AC, BD [II. 1],

∴ AB + BC : BD = AC : AD = constant.

THEOREMS ON HARMONIC SECTION.

Page 387.

1. (i) Let A, P, B, Q be a harmonic range, and S the vertex of the pencil. Through P draw aPb par[l]. to SQ meeting SA, SB at a and b.

Now AP : PB = AQ : QB [Hyp.].

Alternately AP : AQ = PB : QB.

But from the similar △[s] APa, AQS

AP : AQ = aP : SQ;

and from the similar △[s] BPb, BQS

PB : QB = bP : SQ

∴ aP : SQ = bP : SQ,

∴ aP = bP.

Hence, as in Ex. 2, p. 339, it may be shewn that *any* transversal $a'p'b'$ par[l]. to aPb (that is, par[l]. to SQ) has equal parts intercepted by the rays SA, SP, SB.

(ii) Conversely, let the pencil be cut by a transversal $a'p'b'$ par[l]. to SQ, so that $a'p' = b'p'$: then shall the pencil be harmonic.

As before, through P draw aPb par[l]. to $a'p'b'$ (or SQ). Then from the similar △[s] APa, AQS;

AP : AQ = aP : SQ.

And from the similar △[s] BPb; BQS,

PB : QB = bP , SQ.

But since $a'p' = b'p'$ (hyp.). ∴ aP = bP,

∴ aP : SQ = bP : SQ.

Hence AP : AQ = PB : QB.

Alternately AP : PB = AQ : QB,

or, A, P, B, Q is a harmonic range.

Note. As a second converse it may be shewn indirectly that if the range is harmonic, and if in any transversal $a'p' = b'p'$, then $a'p'b'$ is par[l]. to SQ.

2. Let a harmonic pencil be formed by joining a point S to the harmonic range A, P, B, Q; then any transversal shall be cut harmonically by this pencil.

Through P draw *any* transversal aPbq, and also the transversal hPk par^l. to SQ.

Then by Ex. 1 (i), hP = kP.

Hence by Ex. 1 (ii) the range a, P, b, q is harmonic; ∴ any transversal $a'p'b'q'$ par^l. to aPbq is also cut harmonically. [See Ex. 2, p. 339.]

3. (i) In the harmonic pencil {S, APBQ} let one ray SP bisect the angle between the rays SA, SB; then shall SP be perp. to SQ.

Through P draw aPb par^l. to SQ; then since the pencil is harmonic, aP = bP [Ex. 1].

Also in the △aSb, since SP bisects the vert. ∠, and aP = bP,

∴ aS = bS [VI. 3].

Hence the △^{s} SPa, SPb are identically equal, so that ab is perp. to SP.

∴ also SQ is perp. to SP [I. 29].

(ii) *Conversely*, in the harmonic pencil {S, APBQ] let PSQ be a rt. angle; then shall SP, SQ be the internal and external bisectors of the angle ASB.

As before, draw aPb par^l. to SQ, then aP = Pb [Ex. 1], and the ∠^{s} SPa, SPb are rt. angles;

hence the △^{s} SPa, SPb are identically equal [I. 4];

∴ the ∠aSP = the ∠bSP.

That is, SP is the internal bisector of the ∠ASB; and since SQ is perp. to SP, ∴ SQ is the external bisector.

4. Join SQ cutting the transversal $apbq$ in q'.

Then {S, APBQ} is a harmonic pencil by definition; hence, a, p, b, q' is a harmonic range [Ex. 2, p. 387].

but by hypothesis a, p, b, q is a harmonic range,

∴ the points q, q' coincide, since they divide ab externally in the fixed ratio ap : pb.

Hence SQ passes through q, or Qq passes through S.

5. Let Pp, Bb, produced if necessary, meet at S. Join SA, SQ; and let SQ meet the transversal Apb at q'.

Then, as in the last example, $\{S, APBQ\}$ is a harmonic pencil,

$\therefore$ A, p, b, q' is a harmonic range [Ex. 2].

But A, p, b, q is a harmonic range [hyp.];

$\therefore$ q and q' are coincident; or, Qq passes through S;

that is, Pp, Bp, Qq are concurrent.

Similarly it may be shewn that Pq, Bb, Qp are concurrent.

6. *Lemma.* Take two straight lines intersecting at A, and in one of them take *any* two points P, Q, and in the other any two points p, q. Let Pp and Qq intersect at S, and Pq, Qp at S'; now it is proved in Ex. 5, that if B and b are the harmonic conjugates of A with respect to P, Q and p, q, then B, b will lie on the fixed line SS'. Hence it follows, if SS' intersects the given lines at B, b, that A, P, B, Q and A, p, b, q are harmonic ranges.

Now let $PQqp$ be a quadrilateral, and let the sides QP, qp meet at A, and the sides Pp, Qq at S. A *complete quadrilateral* will then be formed.

Let the diagonals Pq and Qp intersect at S': then if SS' meets PQ at B, the range A, P, B, Q is harmonic.

Let the diagonals Pq, Qp meet the third diagonal SA at X and Y: it is required to shew that SA is cut harmonically at X and Y. Join $S'A$.

Then $\{S', APBQ\}$ is a harmonic pencil; therefore it cuts any transversal (such as the third diagonal AS) harmonically. That is, the range A, X, S, Y is harmonic.

Note. The Lemma attached to this proposition furnishes a simple *linear* construction for finding a fourth harmonic to three points.

On Centres of Similarity and Similitude. Page 390.

1. Let ABC be the given △. Take any point P in AC, and draw PQ perp. to BC. From QB cut off QR equal to PQ, and complete the sq. PQRS. Join SC, cutting AB at *s*; and from *s* draw *sp*, *sr* par[l]. to SP, SR to cut AC, BC in *p* and *r*; and from *p* draw *pq* par[l]. to PQ. Then *pqrs* is the required square.

From the similar △[s] CSP, C*sp*, SP : *sp* = SC : *s*C.

From the similar △[s] CSR, C*sr*, SR : *sr* = SC : *s*C;

∴ SP : *sp* = SR : *sr*.

But SP = SR [*constr.*], ∴ *sp* = *sr*.

And since the fig. *pqrs* is by constr. a rectangular par[m]., ∴ it is a square.

2. Let ABC be the △ in which the required △ is to be inscribed: and let X be the △ to which the required △ is to be similar.

In BC, BA take any points P and R respectively, and on PR describe the △PQR equiangular to X, the vertex Q being on the side of PR remote from B.

Join BQ, cutting AC at *q*: and from *q* draw *qp*, *qr* par[l]. respectively to QP, QR, cutting BC, BA at *p* and *r*.

Then *pqr* is the triangle required.

From the similar △[s] BPQ, B*pq*, BP : B*p* = BQ : B*q*.

From the similar △[s] BRQ, B*rq*, BR : B*r* = BQ : B*q*.

∴ BP : B*p* = BR : B*r*;

∴ *pr* is par[l]. to PR.

But by constr. *pq*, *qr* are respectively par[l]. to PQ, QR;

∴ the △*pqr* is similar to the △PQR, that is, to the △X.

3. Let OA, OB be radii of the sector. Join AB, and on AB describe the sq. ABCD, on the side remote from O. Join OD, OC, cutting the arc at *d* and *c*. Then it is clear that *dc* is par[l]. to DC. From *d* and *c* draw *da*, *cb* par[l]. to DA, CB. Join *ab*.

Then as in Examples 1 and 2, it may be shewn that *abcd* is a square.

4. (i) Here A is the external centre of similitude of the two ⊙s whose centres are at I_1 and I,

∴ $I_1A : IA = r_1 : r$ [Ex. 2, Cor. 1, p. 389]
$= I_1D_1 : IX$.

Also I_1D_1 and DIX are par^l. since they are both perp. to BC.

Hence the two △s D_1I_1A, XIA are similar [VI. 6].

∴ the ∠ D_1AI_1 = the ∠ XAI; that is, D_1, X, A are collinear.

(ii) Since BI, BI_1 are the internal and external bisectors of the ∠ ABC, ∴ the pencil {B, $AIYI_1$} is harmonic [p. 385].

5. Taking the fig. and the results of Ex. 33, p. 302, we have from the similar △s ASO, aNO,

SO : NO = SA : Na
= circum.-radius : nine-point-radius.

∴ O is the external centre of similitude of the two circles [p. 389, Ex. 2, Cor. 1].

Again from the similar △s ASG, XNG,

SG : GN = SA : XN
= circum-radius : nine-points-radius.

∴ G is the internal centre of similitude of the two circles.

6. Let C, C′ be the centres of the two fixed ⊙s external to one another, and O the centre of a variable ⊙ touching the others at P, Q respectively. In the fig. taken, the given ⊙s are both external to the ⊙ (O). Then OC, OC′ pass respectively through P and Q [III. 12].

Produce PQ to cut the ⊙ (C′) at P′, and join C′P′.

Then, since OP = OQ, and C′P′ = C′Q,

∴ ∠ OPQ = ∠ OQP = vert. opp. ∠ C′QP = ∠ C′P′Q.
∴ CP and C′P′ are par^l.

∴ P′P passes through the external centre of similitude S.

It will be found that if the given ⊙s are both external, or both internal, to the variable ⊙, then TQ passes through the *external* centre of similitude.

If one of the given ⊙s is within, and the other without the variable ⊙, it will be found that PQ passes through the *internal* centre of similitude.

7. Let C, C′ be the centres of the given circles, and X the given point.

Take S the external centre of similitude, and let C′CS cut the given $\odot^s$ between C, C′ at M and N.

Join SX, and in SX (by describing a $\odot$ through MNX) take a point Y, such that

$$SX \cdot SY = SM \cdot SN.$$

By Ex. 22, p. 254, describe a $\odot$ to pass through X, Y and to touch the $\odot$(C) at P. This $\odot$ shall also touch the $\odot$(C′). Let O be its centre.

Let SP, produced if necessary, meet the $\odot$ (C′) at Q:

then $SX \cdot SY = SM \cdot SN$ [constr.] $= SP \cdot SQ$ [Ex. 2, p. 389].

∴ the $\odot$(O) passes through Q.

It remains to prove that (O) *touches* (C′) at Q, that is, that OQ, C′Q are in one line. Let SPQ meet the $\odot$(C′) again at P′; then since P, P′ are corresponding points, CP is par[l]. to C′P′: hence

$$\angle OQP = \angle OPQ = \text{alt. } \angle C'P'Q = \angle C'QP'$$

but PQP′ is one st. line, ∴ OQC′ is one st. line.

Since two $\odot^s$ can be drawn through X, Y to touch the $\odot$ (C) [Ex. 22, p. 254] it follows that there are two solutions of the problem corresponding to the external centre of similitude. Similarly there will be two more solutions corresponding to the internal centre of similitude.

8. Let A, B, C be the centres of the given $\odot^s$.

Take the general case when the $\odot^s$ are unequal and external to one another. Let (A) be the least of the given $\odot^s$. From centre B, with radius equal to the difference of the radii of (B) and (A) describe a $\odot$; and from centre C with radius equal to the difference between the radii of (C) and (A), describe a $\odot$. Then by the last exercise describe a $\odot$ to pass through A and to touch the two $\odot^s$ of construction. Take O the centre of the last drawn $\odot$, and join OA, cutting the $\odot$ (A) at P. Then a $\odot$ described from centre O with radius OP will touch the three given $\odot^s$. The validity of this construction is apparent at once on drawing the figure.

As each of the given $\odot^s$ may be touched by the required $\odot$ either internally or externally, the required $\odot$ may in general be drawn in $2 \times 2 \times 2$, or 8, ways.

The student will have no difficulty in investigating special cases for himself.

9. Let C_1, C_2, C_3 be the centres of the three $\odot^s$, and r_1, r_2, r_3 their radii. Let S_1 and S_1' be respectively the external and internal centres of similitude of the $\odot^s$ (C_2), (C_3), and let S_2, S_2', S_3, S_3' have corresponding meanings.

(i) $S_1'C_1$, $S_2'C_2$, $S_3'C_3$ shall be concurrent.

Let $S_2'C_2$, $S_3'C_3$ intersect in O; join C_1O and produce it to meet C_2C_3 at X.

The $\triangle^s$ C_2OC_3, C_3OC_1 are on the common base OC_3; hence it may be proved by similar triangles that

the alt. of $\triangle C_2OC_3$: the alt. of $\triangle C_3OC_1 = C_2S_3' : C_1S_3'$

$$\therefore \quad \triangle C_2OC_3 : \triangle C_3OC_1 = C_2S_3' : C_1S_3', \quad = r_2 : r_1.$$

Similarly $\triangle C_1OC_2 : \triangle C_2OC_3 = C_1S_2' : C_3S_2' \quad = r_1 : r_3.$

Ex Æquali, $\triangle C_1OC_2 : \triangle C_3OC_1 = r_2 : r_3.$

But $\triangle C_1OC_2 : \triangle C_3OC_1 = C_2X : C_3X,$

$$\therefore \quad C_2X : C_3X = r_2 : r_3.$$

$\therefore$ X coincides with S_1'; hence $S_1'C_1$, $S_2'C_2$ and $S_3'C_3$ are concurrent.

(ii) To prove S_1, S_2', S_3' collinear.

Join $S_2'S_3'$, and produce it to meet C_2C_3 at Y.

Then C_2C_3, the external diagonal of the quad[l]. $C_1S_2'OS_3'$ is divided harmonically at S_1' and Y [Ex. 6, p. 387]:
hence Y, the harmonic conjugate of S_1' with respect to C_2C_3 is coincident with S_1; or S_1, S_2', S_3' are collinear.

In the same way it may be shewn that each of the ranges of points consisting of one external and two internal centres of similitude are collinear, and also that the three external centres are collinear.

Examples on Pole and Polar.

Page 395.

1. Let A and B be the two given points, and let P be the intersection of their polars. Then by the Reciprocal Property of Pole and Polar, since the polar of A passes through P,

∴ the polar of P passes through A.

Similarly, since the polar of B passes through P,

∴ the polar of P passes through B.

Hence the polar of P passes through both A and B; that is, AB is the polar of P.

2. Let P be the intersection of the given st. lines PQ, PR, and let A and B be their poles.

Then since AB passes through A, ∴ its pole lies on PQ the polar of A.

Similarly since AB passes through B, ∴ its pole lies on PR the polar of B.

Hence the pole of AB is at P, the only point common to PQ and PR.

3. The locus must be the polar of the given point A; for by the Reciprocal Property of Pole and Polar, (i) the pole of any st. line through A must lie on the polar of A; and (ii) any point on the polar of A must be the pole of some st. line through A.

4. Let O be the common centre, P the point of contact of any one of the tangents, and Q its pole: then since the tangent is perp. to OP [III. 18], Q must lie on OP (or OP produced), and OP . OQ = the sq. on the radius of the given circle. But this radius is constant, and OP is constant, ∴ OQ is constant. Hence the locus of Q is a concentric circle.

5. Let PQ be a diameter of one of the ⊙s, and let O be the centre, and r the radius of the other. From O draw OT touching the first ⊙, and join OP cutting the first ⊙ at R. Join QR.

Now OR . OP = OT^2 [III. 36]

$= r^2$, since the circles are orthogonal:

and QRP is a rt. ∠, being in a semicircle.

Hence QR is the polar of P: that is, the polar of P passes through Q.

6. Let P and O be the centres of the two ⊙s which intersect at A, B: and let OP cut AB at Q. Join PA, PB.

Then since the ⊙s are orthogonal, PA and PB touch the ⊙ (O) at A and B: hence OP . OQ = $(\text{radius})^2$ [Ex. 1, page 251].

And OP meets the chord of contact at rt. angles [Ex. 2, p. 196].

∴ AB is the polar of P with regard to the ⊙ (O).

7. Let A and B be the given points, and O the centre of the given ⊙. Then since the polars of A and B are respectively perp. to OA, OB, ∴ one of the ∠s between the polars = the ∠ AOB [Ex. 3, p. 65].

8. Let Q be the point inverse to P with respect to the given ⊙. Draw OY perp. to AB; and through Q draw QX perp. to OP, meeting OY at X.

Then since the ∠s at Q and Y are rt. angles,

∴ the points Q, X, Y, P are concyclic [III. 31].

∴ OX . OY = OP . OQ [III. 36]

$= r^2$ [Hyp.].

But OY is constant, ∴ OX is constant; that is, X is a fixed point.

And since the ∠ OQX is a rt. ∠ [Constr.], ∴ the locus of Q is a circle on OX as diam. [III. 31].

9. Let Q be the point on OP inverse to P, and r the radius of the ⊙ whose centre is O. Draw OX a diam. of the first ⊙. Join PX, and draw QY perp. to OX.

Then OPX is a rt. ∠, being in a semicircle;

and QYX is a rt. ∠ by construction;

∴ the points Q, Y, X, P are concyclic [III. 31];

∴ OX . OY = OP . OQ = r^2.

But since OX is constant, ∴ OY is constant;

hence Y is a fixed point.

Therefore the locus of Q is the st. line perp. to OX through the point inverse to X; that is, the polar of X.

10. Let C and D be the points inverse to A and B respectively, and let AX, BY be the perps. from A and B on the polars of B and A. From A and B draw AN, BN perp. respectively to OB and OA (produced if necessary).

Then OA . OC = OB . OD = r^2 [Definition].

And since the ∠s at M and N are rt. ∠s, the points M, B, N, A are concyclic,

∴ OA . ON = OB . OM [III. 36].

By subtraction

OA . NC = OB . DM.

But NC = BY, and DM = AX [I. 34].

∴ OA . BY = OB . AX.

11. Let RQ cut AD and BC at p and p'. Then it was proved in the solution of Ex. 6. p. 387, that the ranges P, A, p, D and P, B, p', C are harmonic.

Hence by the *harmonic property* of Pole and Polar, the polar of P passes through both p and p': that is, RQ is the polar of P. Similarly it may be shewn that PQ is the polar of R. Hence by the *reciprocal property* of Pole and Polar, PR is the polar of Q; that is to say, the △PQR is self-conjugate with respect to the circle.

12. Let P be the point whose polar with respect to a given circle is to be found.

Through P draw PAD, PBC cutting the ⊙ at A, D and B, C. Let BA, CD intersect at R; and AC, BD at Q. Then, by the last Ex., RQ is the polar of P.

If P is an external point, and RQ cuts the circle at T, T′, then clearly PT, PT′ are the required tangents [Ex. 1, p. 391].

13. Let PQR be a triangle self-conjugate with regard to a circle whose centre is O. Then since QR is the polar of P, ∴ PO is perp. to QR [Def. ii. p. 391].

Similarly RO is perp. to PQ, and consequently QO is perp. to PR [Ex. 19, p. 242]. That is, O is the orthocentre of the △PQR.

14. Let A, P, B, Q be a harmonic range, and O the centre of the given ⊙. Then by the *reciprocal property* of pole and polar, the polars of the points A, P, B, Q are concurrent, since they must all pass through the pole of the line AB. And since these polars are respectively perpendicular to OA, OP, OB, OQ, they must form a pencil whose rays contain severally the same angles as the rays of the pencil {O, APBQ}. But {O, APBQ} is a harmonic pencil [hyp. and def. 2, p. 387], ∴ the pencil formed by the polars is also harmonic.

Examples on Radical Axis and Co-axial Circles.

Page 398.

1. Let TT′ be a common tangent to the two circles, and let their Radical Axis cut TT′ at P. Then, by Definition, the tangential distances of the point P to the two ⊙ˢ are equal: that is, PT = PT′.

2. Let P be any point on the Radical Axis; then the four tangents drawn from P to the two circles are equal [Def.].

Hence a ⊙ described from centre P with any one of these tangents as radius will pass through all four points of contact.

And since the radii drawn from P to the points of contact are also tangents to the given circles, ∴ the ⊙ whose centre is P cuts the given ⊙ˢ orthogonally [p. 240. Def.].

3. As in the last example, all tangents drawn from O to the three ⊙ˢ are equal, ∴ a circle from centre O with radius OT will pass through all the points of contact. And since the radii of this ⊙ drawn to the points of contact are also tangents to the given ⊙ˢ, ∴ the ⊙ whose centre is O cuts the given ⊙ˢ orthogonally.

4. Let the ⊙s (A), (B), (C) touch one another two and two, and let OT, OT′ be the common tangents of the ⊙s (A), (B) and (A), (C) at their points of contact.

Then since OT and OT′ are tangents to the ⊙ (A),

∴ OT = OT′.

That is, tangents drawn from O to the ⊙s (B), (C) are equal:

∴ O is a point on the radical axis of the ⊙s (B), (C).

But the radical axis of two ⊙s which touch one another is clearly the common tangent at their point of contact.

Hence the common tangent to the ⊙s (B), (C) passes also through O.

5. Take the figure of p. 243.

Since the ∠s BEA, BEC are rt. angles, ∴ ⊙s described on AB, BC as diams. pass through E [III. 31];

that is, BE is the common chord of the ⊙s on AB and BC.

Similarly AD and CF are respectively the common chords of the ⊙s on AB, AC and on BC, CA.

Hence O, the point of intersection of the common chords, is the radical centre [p. 397. Cor.].

6. See solution of Ex. 7, p. 252. Observe that the required point B is the *inverse* of the given point A with regard to the given circle.

7. Since by the last Example all ⊙s which pass through the fixed point A and cut a given ⊙ orthogonally, pass also through a second fixed point B (the inverse of A with regard to the given ⊙), ∴ the locus of their centres is the st. line bisecting AB at rt. angles.

To find this point B, draw *any* radius CT to the given ⊙:

describe a ⊙ to pass through A and touch CT at T [Ex. 28, p. 238].

Let this ⊙ cut CA at B. Then B is the required point; for

CA . CB = CT^2 [III. 36].

8. Let C be the centre of the given ⊙, and A, D the given points. Now by Ex. 6 all ⊙s through A cutting the given ⊙ orthogonally must pass through B the inverse point of A with respect to the given ⊙.

Determine B as in the last Example. Then the ⊙ circumscribed about the △ABD is that required.

9. Let P be the centre of any ⊙ which cuts the two given ⊙s orthogonally at T and T′.

Then PT = PT′, being radii.

Also PT and PT′ are tangents to the given ⊙s, since the ⊙s are cut orthogonally.

Hence the locus of P is the radical axis of the two given ⊙s.

10. Let C and C′ be the centres of the given ⊙s, and A the given point.

Then all ⊙s through A cutting the ⊙ (C) orthogonally pass through B the inverse of A with respect to the ⊙ (C);

and all ⊙s through A cutting the ⊙ (C′) orthogonally pass through B′ the inverse of A with respect to the ⊙ (C′).

Determine the points B and B′ as in the solution to Ex. 7.

Then the ⊙ about the △ABB′ is that required.

Note that by Ex. 9 the centre of this ⊙ is on the radical axis of the given ⊙s(C) and (C′).

11. Let A, B be the centres of the two given ⊙s; PQ, PR tangents to them from the given point P. Let the Radical Axis cut AB at S.

Draw PM, PN perp. respectively to AB and the Radical Axis; and bisect AB at O.

Then $AP^2 - BP^2 = 2AB \,.\, OM$ [Ex. 8, p. 159].

And $AQ^2 - BR^2 = AS^2 - SB^2$ [Ex. 1, p. v.]

$= 2AB \,.\, OS$ [Ex. 8, p. 159],

∴, by subtraction,

$$AP^2 - AQ^2 - (BP^2 - BR^2) = 2AB\,(OM - OS),$$

or

$$PQ^2 - PR^2 = 2AB \,.\, SM = 2AB \,.\, PN.$$

12. Let A, B be the centres of two ⊙ˢ of the system, and let their Radical Axis cut AB at S. From P, *any* point in the Radical Axis, draw tangents PQ, PR to the two ⊙ˢ; then $PQ = PR$ [Hyp.]. From centre P, with radius PQ, describe a ⊙ cutting AB at L, L′. Then L, L′ shall be fixed points for all positions of P.

From S draw tangents ST, ST′ to the two ⊙ˢ.

Then

$$\begin{aligned} SL^2 &= PL^2 - PS^2 \quad [\text{I. } 47] \\ &= PQ^2 - PS^2 \\ &= PA^2 - QA^2 - PS^2 \\ &= PS^2 + AS^2 - QA^2 - PS^2 \\ &= AS^2 - AT^2 \\ &= ST^2. \end{aligned}$$

But ST is independent of the position of P; ∴ L is a fixed point.

Similarly $$SL' = ST' = ST = SL.$$

13. Let the radical axis cut the line of centres at S, and let *any* ⊙ of the system cut the same line at XY. If ST is the tangent from S to this circle, then by definition $ST = SL = SL'$, where L, L′ are the limiting points.

Also $$SL^2 = ST^2 = SX \,.\, SY \quad [\text{III. } 36],$$

∴ L, X, L′, Y form a harmonic range [Ex. 2, p. 385], since S bisects LL′.

14. With the notation of the last Ex., since L, X, L′, Y form a harmonic range, ∴ the polar of L with regard to any circle of the system which cuts the line of centres at X, Y, must cut this line perpendicularly at L′ [Ex. 4, p. 394]. But L′ is a *fixed* point [Ex. 12]; ∴ the polar of L for all circles of the system is the same.

15. Let O, O′ be the centres of two ⊙ˢ which cut one another orthogonally at T. Let AB, a diameter of the ⊙ (O), cut the ⊙ (O′) at P, Q.

Then $$OP \,.\, OQ = OT^2 = OB^2,$$

∴ A, P, B, Q form a harmonic range [Ex. 2, p. 385].

ON TRANSVERSALS. **Page 402.**

1. Take the figure and the results of p. 297,

We have, since AF = AE, BF = BD, CD = CE,

$$\frac{AF}{FB} \cdot \frac{BD}{DC} \cdot \frac{CE}{EA} = 1;$$

∴ AD, BE, CF are concurrent [Ex. 1, p. 400].

2. Let the four st. lines EAB, EDC, FDA, FCB form the complete quad[l] ABCDEF; and let X, Y, Z be the middle points of the diagonals BD, AC, EF.

Then shall X, Y, Z be collinear.

Take P, Q, R the middle points of EA, ED, AD.

Then from the △AEC, since P and Y are the middle points of AE, AC,

∴ PY is par[l] to EC, and cuts AD at its middle point R.

Similarly PZ is par[l] to AF, and cuts ED at its middle point Q; also QX is par[l] to EB, and cuts AD at R.

Hence QX, XR, PZ, ZQ, RY, YP are respectively halves of EB, BA, AF, FD, DC, CE.

But the sides of the △EAD are cut by the transversal BCF,

$$\therefore \frac{EB}{BA} \cdot \frac{AF}{FD} \cdot \frac{DC}{CE} = 1.$$

Hence $$\frac{QX}{XR} \cdot \frac{PZ}{ZQ} \cdot \frac{RY}{YP} = 1.$$

∴ X, Y, Z, points in the sides of the △PQR, are collinear.

[See Rouché et de Comberousse, *Traité de Géométrie*, p. 205.]

3. Let the △s ABC, A′B′C′ be co-polar; that is, let AA′, BB′, CC′ meet at S: then shall they be co-axial; that is X, Y, Z the intersections of BC, B′C′, of CA, C′A′ and of AB, A′B′ shall be collinear.

From the △SAB and the transversal A′B′Z,

$$\frac{AZ}{ZC} \cdot \frac{BB'}{B'S} \cdot \frac{SA'}{A'A} = 1.$$

From the $\triangle$ SBC and the transversal B'C'X,

$$\frac{B'S}{BB'} \cdot \frac{C'C}{SC'} \cdot \frac{XB}{CX} = 1.$$

From the $\triangle$ SCA and the transversal C'A'Y,

$$\frac{AA'}{A'S} \cdot \frac{SC'}{C'C} \cdot \frac{CY}{YA} = 1.$$

Multiplying these results we have

$$\frac{AZ}{ZB} \cdot \frac{XB}{CX} \cdot \frac{CY}{YA} = 1.$$

$\therefore$ X, Y, Z are collinear.

Conversely, let X, Y, Z be collinear; then shall AA', BB', CC' be concurrent.

Let BB', CC' meet at S.

Then the $\triangle^s$ BZB', CYC' are co-polar; hence by the first proof they are co-axial; that is, A, A', S are collinear,

or AA', BB', CC' meet at S.

4. Let C_1, C_2, C_3 be the centres of the three $\odot^s$, and r_1, r_2, r_3 their radii. Let S_1 and S_1' be respectively the external and internal centres of similitude of the $\odot^s$ (C_2), (C_3), and let S_2, S_2', S_3, S_3' have corresponding meanings.

To prove S_1', S_2' and S_3 collinear.

By definition we have

$$\frac{C_1S_3}{S_3C_2} = \frac{r_1}{r_2}; \quad \frac{C_2S_1'}{S_1'C_3} = \frac{r_2}{r_3}; \quad \frac{C_3S_2'}{S_2'C_1} = \frac{r_3}{r_1}.$$

$$\therefore \frac{C_1S_3}{S_3C_2} \cdot \frac{C_2S_1'}{S_1'C_3} \cdot \frac{C_3S_2'}{S_2'C_1} = 1.$$

Hence from the $\triangle C_1C_2C_3$, the points S_1', S_2', S_3 are collinear [p. 401, Ex. 2, Converse].

In the same way it may be shewn that each of the ranges of points consisting of one external and two internal centres of similitude are collinear, and also that the three external centres are collinear.

Miscellaneous Examples on Book VI.

Page 403.

1. By par^{ls}., BF : FA = BD : DC = AE : EC.

And $\triangle$BFD : $\triangle$AFE = BF : FA [VI. 1].

And $\triangle$AFE : $\triangle$CDE = AE : EC.

$\therefore$ $\triangle$BFD : $\triangle$AFE = $\triangle$AFE : $\triangle$CDE.

2. Let $\angle$ABC = $\angle$DEF; and $\angle$ACB = supplement of $\angle$DFE. With centre A and radius AC describe a $\odot$ to cut BC in C′. Then AC = AC′, $\therefore$ $\angle$BC′A = supplement of $\angle$BCA = $\angle$DFE. $\therefore$ $\triangle^s$ DEF, ABC′ are equiangular;

$\therefore$ ED : DF = BA : AC′ = BA : AC [VI. 4].

3. The diameters of the $\odot^s$ about ABE, ACE are in the ratio of AB to AC [Ex. 5, p. 383]. But, because AE bisects $\angle$BAC,

$\therefore$ AB : AC = BE : EC [VI. 3].

4. Let A and B be the two other fixed pts. Divide AB at C in the given ratio. Join OC: and, through O, draw MON perp. to OC. MN is the required line. [Ex. 2, p. 329].

5. Let AB > AC. Draw CF perp. to AD. And let CF produced cut AB in M. Then $\triangle^s$ CAF, MAF are identically equal, $\therefore$ CF = MF; and AM = AC. $\therefore$ BM = difference of sides AB, AC. Bisect BM in K. Then AK = $\frac{1}{2}$sum of sides AB, AC. Join KF, XF. Then K, F, X being the middle pts. of the sides of BMC, KF is par^l. to BC and XF is par^l. to BA. $\therefore$ by similar $\triangle^s$ DXF, FKA,

XD : KF = XF : KA;

that is, XD : BX = BK : KA

= $\frac{1}{2}$ diff. of sides : $\frac{1}{2}$ sum of sides.

6. BD : DC = BA : AC = BE : EC [VI. 3 and A].

$\therefore$ BD − DC : BD + DC = BE − EC : BE + EC [V. 13],

or, 2OD : 2OB = 2OB : 2OE.

Hence OB is a mean proportional between OD and OE

7. Draw PX perp. to AB, and QY perp. to CD. Then by similar △ˢ MPX, NQY,

PM : QN = PX : QY = constant.

Let MN meet PQ in O.

Then OP : OQ = PM : QN = constant

Hence MN passes through a fixed pt. O, which divides PQ in the ratio PX to QY (*internally* or *externally* according as PX is in the opposite or in the same direction as QY).

8. Because C bisects arc AB, ∴ chord AC = chord BC.

But rect. AD, BC + rect. DB, AC = rect. AB, DC [Prop. D].

∴ rect. AC (AD + DB) = rect. AB, DC ;

∴ AD + DB : DC = AB : AC [VI. 16].

9. Because CD bisects ∠ ACB *internally*,

∴ BD : DA = BC : CA = 1 : 2.

And because CE bisects ∠ ACB *externally*,

∴ BE : EA = 1 : 2.

Hence AD = 2DB, AB = 3DB, DE = 4DB.

Also △ DCB : △ ACD : △ ACB : △ DCE

= DB : AD : AB : DE

= 1 : 2 : 3 : 4.

10. Because DE is par[l]. to the tangent at A, ∴ it makes with AB, AC angles respectively equal to ACB, ABC; [III. 32];

or ∠ ADE = ∠ ACB and ∠ AED = ∠ ABC.

∴ △ˢ ABC, AED are equiangular ;

∴ AB : AC = AE : AD ;

∴ rect. AB, AD = rect. AC, AE [VI. 16].

11. Let △ ABC be right ∠ᵈ at A. From D in BC, draw DE, DF perps. on AC, AB. Then the rt. ∠ᵈ △ˢ BFD, DEC are similar.

∴ rect. BD, DC = rect. BF, DE + rect. DF, CE [VI. 31].

But DE = FA and DF = EA.

∴ rect. BD, DC = rect. BF, FA + rect. CE, EA.

12. Let BD, CE cut in O. Because BO : OD = CO : OE,

∴ DE is par^{l}. to BC, and △^{s} BOC, DOE are similar;

∴ BC : DE = BO : OD

= 4 : 1.

But BA : EA = BC : ED

= 4 : 1.

∴ BA − EA : EA = 4 − 1 : 1 [v. 13],

or BA : EA = 3 : 1.

13. Let P, Q be two fixed pts., AB any st. line between them. Draw PM, QN perps. on AB, and let PQ cut MN in O.

Then OP : OQ = PM : QN = constant.

∴ MN always passes through the fixed pt. O, which divides PQ internally in the given constant ratio.

14. Because ∠ PAC = ∠ ADC [III. 32], ∴ △^{s} PAD, PCA are equiangular [I. 32];

∴ AD : CA = PA : PC.

Similarly, BD : CB = PB : PC.

But PA = PB. ∴ AD : CA = BD : CB.

∴ rect. AD, CB = rect. CA, BD.

15. Because ∠ DAC = ∠ ABD, ∴ △^{s} DAC, DBA are equiangular.

∴ DC : DA = AC : BA

= circum-diam. of ACD : circum-diam. of ABD

[Ex. 5, p. 383].

16. Let F be between E and B. Then ∠ EFC = ∠ EDC in same segment = complement of ECD = ∠ CGH.

∴ △^{s} CFE, CGH are equiangular [I. 32].

∴ CE : EF = CH : HG.

∴ rect. CE, HG = rect. CH, EF.

17. Make ∠ CAD = ∠ ABC. Then △^{s} BDA, ADC are similar.

∴ BD : DA = DA : DC.

∴ DA is a mean proportional between BD and DC.

18. The common tangent at O makes with OA an ∠ equal to ∠ OQP and to ∠ OBA [III. 32]. ∴ ∠ OQP = ∠ OBA. ∴ PQ is par[l]. to AB. ∴ ∠ PQC = alt. ∠ QCB = ∠ CPQ in alternate segment. ∴ chord CQ = chord CP. ∴ OC bisects ∠ BOA [III. 28, 27].

∴ OP : OQ = OA : OB = AC : BC [VI. 3].

19. Taking the figure in which D is on the side of O remote from AB, the ∠ CEO = comp[t]. of ∠ B = comp[t]. of $\frac{1}{2}$ ∠ COA at centre = ∠ OCD. ∴ △[s] ODC, OCE are equiangular.

∴ OD : OC = OC : OE.

∴ rect. OD, OE = sq. on OC.

20. Join AD, BD. Then the ∠ BDY = ∠ BAD = ∠ BDX. ∴ DB bisects ∠ YDX internally. Again, DA is perp. to DB,

∴ DA bisects ∠ YDX externally.

∴ XB : BY = XD : DY = XA : AY [VI. 3 and A].

∴ BX : AX = BY : AY.

21. Let P, Q be the given pts. Divide PQ, internally and externally, at A and B in the given ratio [Ex. 1, p. 384]. On AB as diameter describe a ⊙. Then the distances of P and Q from any pt. on this circle are in the given ratio [Ex. 4, p. 386]. The pt. or pts., if any, where this ⊙ cuts the given ⊙ are the pts. required.

22 Produce AO to meet the ○[ce] at B, and let LP produced meet OA at Z.

Join LA, LB.

Then since the arc AP = arc AQ,

∴ LA bisects the ∠ VLZ internally

and since LB is perp. to LA [III. 31],

∴ LB bisects the ∠ VLZ externally.

Hence Z divides BA externally in the fixed ratio BV : VA [p. 385].

23. Let O, O′ be the centres. Then, because BE and C′O′ are both at rt. $\angle^s$ to ABC′,

$$\therefore AB : BC' = AE : EO' \text{ [VI. 2].}$$

$$\therefore AB : 2BC' = AE : EA'.$$

Similarly $\quad A'B' : 2B'C = A'E : EA.$

$$\therefore AB : 2BC' = 2B'C : A'B'.$$

$\therefore$ rect. AB, A′B′ = four times rect. BC′, B′C.

24. Let A, B be the centres of the fixed $\odot^s$; and C the centre of the circle touching them externally in D and E respectively. Let DE cut AB in S and the $\odot$ B again in E′. Join BE′. Then D is in AC, and E in BC. Because CD = CE, and BE = BE′,

$\therefore \angle CDE = \angle CED = \angle BEE' = \angle BE'E.$ $\therefore$ BE′ is par[l]. to AD.

$$\therefore AS : BS = AD : BE'.$$

That is, S is the external centre of similitude of the fixed $\odot^s$.

25. Because DC bisects $\angle$ ADB,

$\therefore DA : DB = CA : CB = AE : BF.$ And the $\angle^s$ AED, BFD are rt. $\angle^s$, $\therefore$ the $\triangle^s$ AED, BFD are similar [VI. 7. Cor.].

$\therefore$ rect. DA, DB = rect. DE, DF + rect. AE, BF [VI. 31].

But rect. DA, DB = rect. AC, BC + sq. on DC [VI. B]

= rect. AE . BF + sq. on DC;

$\therefore$ rect. DE, DF = sq. on DC.

26. Let D be middle pt. of BC, AX parallel to BC: and let DX cut AB in Y and AC in Z. Then, by similar $\triangle^s$ XYA, DYB,

$$XY : DY = XA : DB.$$

And, by similar $\triangle^s$ XZA, DZC,

$$XZ : DZ = XA : DC.$$

But $\quad DB = DC. \quad \therefore XY : DY = XZ : DZ:$

that is, XD is divided harmonically at Y, Z. [Def. p. 385.]

27. Let the line cut the median in X. Through X draw EF par[l]. to the base. Then EF is bisected in X. Hence, by the last example, the line is divided harmonically.

28. Let ∠ BAC be bisected internally and externally by AX and AY, and let the four concurrent lines be met by a fifth line at B, X, C, Y.

Then BX : XC = BA : AC = BY : YC [VI. 3 and A].

∴ BC is cut harmonically, at X, Y.

29. See Ex. 3, p. 387.

30. Divide AB at G, so that AB is to AG in the given ratio [Ex. 1, p. 346]. Join CG, and produce it to meet AE at E. Then, because EA is par[l]. to BC,

CE : GE = AB : AG = given ratio [VI. 2].

31. Produce PA to X, so that PA : PX = given ratio [VI. 12].

Draw XR par[l]. to AB cutting AC in R, and let PR cut AB in Q.

Then PQ : PR = PA : PX = given ratio.

32. Let P be the given pt. within the circle ABD. Through P draw the diam. APB, and on it take AP : PC in the given ratio. With P as centre and radius equal to a mean prop[l]. between BP and PC describe a circle cutting ADB in D (or D′); join DP and produce it to E; then DE is the required chord.

For, by construction BP : PD = PD : PC,

and since the rect. PE, PD = the rect. PB, PA [III. 35],

∴ BP : PD = PE : PA [VI. 16];

∴ PE : PA = PD : PC,

∴, *alternately*, PE : PD = AP : PC

= the given ratio.

Since the circle of construction will in general cut the given circle in two points there will be two solutions.

33. Let A be the common pt. of contact, and B the pt. on the common tangent BA. Let a ⊙ having centre B, cut one of the ⊙s in C, and let BC cut this ⊙ again in D.

Then sq. on BA = rect. BC, BD

= sq. on BC + rect. BC, DC.

But BA and BC are constant. ∴ DC is constant.

34. Let SPT meet CA, CB produced in S and T. Draw PN perp. to BC.

Then △SCT : △ACB = CS . CT : CA . CB

$= CS . CT : CP^2$.

Also CS : CP = CP : CM,

and CT : CP = CP : CN [VI. 8];

∴ $CS . CT : CP^2 = CP^2 : CM . CN$

$= CP^2 : CM . MP$

= CA . CB : CM . MP

∴ △SCT : △ACB = CA . CB : CM . MP

= △ACB : △CMP.

35. The tangents at B and C make with BC angles equal to ∠BAC in alt. segment. And AD, AE being parl. to these,

∠BDA = ∠AED = ∠A. ∴ AD = AE;

again the △s BDA, AEC being each similar to BAC are similar to one another, BA, AC being homologous sides.

∴ BD : CE = △BDA : △CAE [VI. 1]

= dupl. ratio of BA : AC.

36. Let X, Y be the centres of the ⊙s on AE, EB.

AE + EB = 2OB = 4EB;

∴ AE = 3EB; or PX = 3QY.

Now XL : YL = PX : QY = 3 : 1.

∴ XY = 2YL.

But XY = XE + EY = PX + QY = 4QY = 4BY,

∴ YL = 2BY.

∴ BL = BY.

37. Because rect. AC, CB = rect. CD, CE,

∴ AC : CE = CD : CB;

∴ △ˢ ACE, DCB are similar [VI. 6],

and the pts. A, C, B, E are concyclic.

And since AB is fixed, and the ∠ ACB is constant, ∴ the ⊙ ACB is fixed.

But the ∠ˢ ACE, BCE are equal;

∴ E bisects the fixed arc AEB.

38. By IV. 10, the △ˢ ABC, BEC are similar

Also AE = BC = BE.

∴ AB : BC = AE : EC;

AB + BC : BC = AC : EC [V. 13]

= △ABC : △BEC

= △ABC : △ADE.

∴ AB : BC = fig. DBCE : △ADE [V. 13].

39. Let H, K be the centres of the equal ⊙ˢ; G that of the inscribed ⊙, which touches the equal ⊙ˢ in E and F and the outer ⊙ in D. Then G, E, H and G, F, K, and D, G, C, are collinear.

Produce GEH to meet circle (H) in L;

then rect. LG, GE = sq. on GC;

∴ EG : GC = GC : LG [VI. 17]

= EG + GC : GC + LG [V. 12]

= CD : LE + DC

= 1 : 2.

∴ EG : EG + GC = 1 : 3;

that is DG : DC = 1 : 3;

∴ 2DG : DC = 2 : 3.

40. Let the quad^l. ABCD touch the ⊙ at G, H, K, M; and let DA, CB meet at L. Because LE = LF and LM = LH, and OM = OH, ∴ △^{s} OEM, OFH are identically equal. And ∠ MOA = ∠ GOA and ∠ HOB = ∠ GOB.

∴ ∠^{s} EOA, BOH together = half ∠^{s} EOG, FOG = a rt. ∠.

∴ ∠ EOA = complement of ∠ BOH = ∠ OBH.

∴ △^{s} EOA, FBO are equiangular [I. 32].

∴ AE : OE = OF : BF.

∴ rect. AE, BF = rect. OE, OF.

Similarly rect. DE, CF = rect. OE, OF.

∴ AE : DE = CF : BF [VI. 16].

41. Considering the △ABC as the limiting form of a quadrilateral AFBC touching the ⊙, it follows by last example that

BX : XF = AY : YC,

for F is the pt. where the tangent from B cuts the tangent from A.

42 and **43.** Let AB be the base of the segment. Bisect AB in C. Draw CD perp. and equal to AB, on the same side of AB as is the segment. Draw CE par^l. to AD cutting the arc in E: and draw EF perp. to AB. EF shall be the side of the square inscribed in the segment. For, by similar △^{s} ACD, CFE, since DC = twice AC, ∴ EF = twice CF.

44. Let ABC be the isosceles △. Draw AD perp. to the base BC. At A make the ∠^{s} DAE, DAF each = $\frac{1}{3}$ of a rt. ∠, E and F being in BC. Then AEF is an equilateral △. From AE, AF cut off AG, AH each equal to the mean proportional between AE and BC. Then by similar △^{s} AGH, AEF,

△AEF : △AGH = dup. ratio of AE : AG [VI. 19]

= AE : BC

= EF : BC

= △AEF : ABC.

∴ the equilateral △AGH = given △ABC.

45. Let AB be the given difference. Draw BC at rt. ∠^{s} and equal to AB. Produce AC to D making CD = BC = AB. AD is a side of the square required.

46. With the given diameter EB describe a ⊙ EABC. Make ∠ BEC = given vertical ∠. Divide BC in given ratio at D. Bisect arc BC in F. Produce FD to A. ABC shall be required △. For ∠^{s} BAC, BEC in same segment are equal, and since

arc BF = arc CF, ∴ ∠ BAF = ∠ CAF.

∴ AB : AC = BD : CD = given ratio.

47. Let AD be the given median. Produce AD to E, making DE = AD. On AE describe segment of ⊙ ABE containing an angle equal to the supplement of given vertical ∠. Draw the base BDC making required ∠ with the median AD, cutting the arc ABE in B, and making DC = BD. ABC shall be the required △. For, because BC, AE bisect one another, ABEC is a par^{m}. ∴ ∠ BAC = supplement of ∠ ABE.

48. Let XY be the given st. line, and P, Q the given pts. Join PQ and in it take a pt. F so that rect. PF, PQ = the rectangle contained by the segments of any chord of the circle through P [VI. 12]. Let QP and YX be produced to meet at Z. Let K be the length of a chord of the ⊙ which subtends at the ○^{ce} an angle equal to ∠ QZY; through F draw a line FBD cutting off a chord BD equal to K [Ex. 9, p. 197]. Draw PBA meeting ⊙ in B, A, and join AQ meeting the ⊙ in C. Then ABC shall be the required △.

Because rect. PF, PQ = rect. PB, PA;

∴ PF : PB = PA : PQ,

∴ △^{s} PBF, PAQ are similar [VI. 6].

∴ ∠ PFB = ∠ PAC

= ∠ BDC, (or the supplement of BDC ;)

∴ DC is par^l. to PQ.

And because ∠ DCB = ∠ QZY;

∴ BC is par^l. to XY.

49. Let P, Q, R be the given pts. Join PQ and determine a pt. F in it as in Ex. 48.

In the circle inscribe a △DBC so that DB and BC pass through F and R respectively, while DC is par^l. to PQ [Ex. 48].

Produce PB to meet the ○^{ce} in A; join QA meeting the ○^{ce} in C′ and join DC′.

Then ∠BAQ = ∠FDC′ in the same segment.

Also, as in Ex. 48, the △^{s} PFB, PAQ are similar;

∴ ∠PAQ = ∠PFB = alt. ∠FDC.

∴ ∠FDC = ∠FDC′.

Hence C′ coincides with C, and the △ABC fulfils the required conditions.

50. Take the case in which the points are in the following order: O, A, B, X, Y.

Take OE a mean prop^l. between OA and OY and describe a ⊙ with O as centre and OE as radius. Take P on the ○^{ce} of this ⊙; describe a ⊙ round PAY and also round PBX. Then OP touches each of these ⊙^{s}, since $OP^2 = OA \cdot OY = OB \cdot OX$.

∴ ∠OPB = ∠PXB [III. 32].

But ∠OPB = sum of ∠^{s} OPA, APB,

and ∠PXB = sum of ∠^{s} XPY, PYA [I. 32],

∴ sum of ∠^{s} OPA, APB = sum of ∠^{s} XPY, PYA;

but ∠OPA = ∠PYA [III. 32]:

∴ ∠APB = ∠XPY.

51. Through Q draw a st. line par^l. to the given st. line. This is the required locus.

52. Let C be the centre of the given ⊙. In OC take D, so that OD : OC = given ratio. Then △^{s} OPC, OQD are similar, and DQ : CP = given ratio. But CP is constant, and D is fixed; ∴ locus of Q is a ⊙, having centre D and radius DQ.

53. Let O be a given pt. Take OP : OQ = given ratio, where P is on given line and ∠POQ = given ∠.

Take any other pt. P′ on given line, and make ∠P′OQ′ = ∠POQ, and ∠OQQ′ = ∠OPP′. Then, because ∠P′OQ′ = ∠POQ and ∠P′OQ is common, ∴ ∠ POP′ = ∠ QOQ′. ∴ △s POP′, QOQ′ are similar.

∴ OP′ : OQ′ = OP : OQ = given ratio.

∴ Locus of extremity Q′ is the st. line through Q making with OQ the same ∠ that the given st. line makes with OP.

54. Let E be middle pt. of AB. Then, since diagonals of a par^{m}. bisect one another, E is middle pt. of CD.

Draw DO, par^{l}. to EP, meeting CP in O. Then △s DOC, EPC are similar. ∴ OC = twice PC, and OD = twice PE; hence O is a fixed point, and OD is of constant length. ∴ the locus of D is a ⊙, having the fixed pt. O as centre.

55. See Ex. 4, p. 386.

56. Let A, B be the centres of the given ⊙s: and let O be a pt. from which the ⊙s subtend equal ∠s.

Let OS, OS′ and OT, OT′ be tangents to the ⊙s from O.

Then ∠s SOS′, TOT′ are equal; ∴ the ∠s SOA, TOB are equal. And ∠s ASO, BTO are rt. ∠s. ∴ △s SAO, TBO are equiangular.

∴ OA : OB = SA : TB = fixed ratio.

∴ locus of O is a ⊙. [Ex. 55.]

57. Let OA, OB be the two given lines. Produce AO to A′, and in OA′ and OB take points H and K, so that OK : OH = the given ratio.

Draw OC par^{l}. to HK. OC is the required locus.

For, draw KQ perp. to OB, and QR perp. to OA. Also, from any pt. P in OC, draw PM perp. to OA and PN perp. to OB.

Then PM : PN = QR : QK.

But △s OHQ, OKQ on same base OQ and between same par^{ls}. are equal.

∴ rect. QR, OH = rect. QK, OK.

∴ QR : QK = OK : OH.

∴ PM : PN = given ratio.

58. Because TP is par[l]. to T′P′,

∴ ∠ ST′P′ = ∠ STP = ∠ PQT [III. 32] = supplement of ∠ P′QT.

∴ Q, T, T′, P′ are concyclic. If then TQ, T′P′ cut in X, the rect. XQ, XT = rect. XP′, XT′. ∴ tangents from X to the ⊙s are equal. ∴ X is on the radical axis.

59. Let D, E, F be the vertices of the equilateral △s. Then the △s BAE, FAC are identically equal. ∴ BE = FC. But the △s BZA, CYA are similar.

∴ ZA : AB = YA : AC.

∴ ZA : YA = AF : AC.

But ∠ ZAY = ∠ FAC. ∴ △s ZAY, FAC are similar.

∴ ZY : CF = AY : AC.

Similarly XY : BE = AY : AC.

Hence XY = YZ = ZX.

60. Let ABC be the triangle; S, I the centres, and R, r the radii of the circumscribed and inscribed circles.

(i) To prove $SI^2 = R^2 - 2Rr$.

Join AI, and produce it to meet the ○ce of the circum-⊙ at X. Join XS, and produce it to meet the ○ce again at Y. Join XC, and draw IE perp. to AC. Join YC.

Then in the △s IAE, XYC,

∠ IAE = ∠ XYC [III. 21]; and ∠ IEA = ∠ XCY [III. 31];

hence the △s IAE, XYC are equiangular [I. 32],

∴ IE : IA = XC : XY [VI. 4],

∴ IE . XY = IA . XC [VI. 16].

But IE = r; XY = 2R; and XC = XI [Ex. 16, p. 383],

∴ $2Rr$ = XI . IA.

Join SI, and produce it both ways to meet the ○ce at P, Q.

Hence XI . IA = PI . IQ [III. 35]

= (PS + SI) (SQ − SI)

$= R^2 - SI^2$,

or, $SI^2 = R^2 - 2Rr$.

Similarly, if I_1, I_2, I_3 are the centres and r_1, r_2, r_3 the radii of the escribed ⊙s, it may be shewn that

$$SI_1^2 = R^2 + 2Rr_1; \quad SI_2^2 = R^2 + 2Rr_2;$$

and

$$SI_3^2 = R^2 + 2Rr_3.$$

(ii) To prove $IN = \frac{R}{2} - r$. (Feuerbach's Theorem.)

Several proofs of this theorem have been given, those depending upon pure geometry being difficult and complicated. [See Casey's *Sequel to Euclid*, p. 105, Milne's *Companion to Weekly Problem Papers*, Chapter VI., p. 185.]

We here give an outline of Feuerbach's proof, one step of which depends on trigonometrical work.

Let S, I, and N be the centres of the circumscribed, inscribed, and nine-point ⊙s of the △ABC, and O its orthocentre. Let AO meet BC at D, and the Oce of the circumscribed ⊙ at G. Join SI, IO, and SO; and let SO produced both ways meet the Oce at P and Q.

Then N is the middle point of SO [Ex. 33, p. 302].

And since IN is a median of the △SIO,

$$\therefore SI^2 + IO^2 = 2IN^2 + 2SN^2,$$

or

$$SI^2 + IO^2 = 2IN^2 + \tfrac{1}{2}SO^2 \quad \text{.................(i)}$$

But $SI^2 = R^2 - 2Rr$; and $SO^2 = R^2 - PO \cdot OQ$ [II. 5]

$$= R^2 - AO \cdot OG.$$

Also it may be proved by trigonometry from the △IAO that

$$IO^2 = 2r^2 - AO \cdot OD$$

$$= 2r^2 - \tfrac{1}{2}AO \cdot OG \quad \text{[Ex. 21, p. 244].}$$

Substituting these results in (i), we have

$$2(R^2 - 2Rr) + 4r^2 - AO \cdot OG = 4IN^2 + R^2 - AO \cdot OG,$$

or,

$$R^2 - 4Rr + 4r^2 = 4IN^2,$$

i.e.,

$$(R - 2r)^2 = (2IN)^2,$$

$$\therefore \frac{R}{2} - r = IN.$$

Remembering that the radius of the nine-points-circle is half the radius of the circum-⊙, we see that the nine-points ⊙ *touches* the inscribed ⊙.

Similarly it may be shewn that

$$NI_1 = \frac{R}{2} + r_1; \text{ \&c.},$$

so that the nine-points-⊙ touches also the three escribed ⊙s.

BOOK XI.

EXERCISES.

Page 429.

1. Let AB be the perpendicular, and AP any other st. line drawn from the external point A to the plane XY.

Join BP; then by Def. I. p. 384, AB is perp. to BP.

Hence the ∠ ABP > the ∠ APB; ∴ AP > AB. [I. 99.]

2. Let AP, AQ be equal st. lines drawn from A to the plane XY, and let AB be the perp. drawn from A to that plane.

Then by Def. I. p. 410, BP, BQ are at rt. angles to AB.

Hence the △s ABP, ABQ are identically equal.

[Ex. 12, p. 99.]

∴ the ∠ PAB = the ∠ PAQ.

3. Place the spirit-level along any two intersecting lines BP, BQ in the plane. Then if these lines are found to be horizontal, a vertical line AB is perp. to both, and therefore [XI. 4] perp. to the plane XY in which they are: that is, the plane XY is horizontal.

Consider the inclined plane BC in the fig. to Def. 7, p. 412; and let AB be its common section with the horizontal plane AD. Then AB is horizontal, since it lies in a horizontal plane. Hence all st. lines drawn in the plane BC par^l. to AB are also horizontal. If therefore two *par^l*. lines are shewn by the spirit-level to be horizontal, it cannot be inferred that the plane in which they are is horizontal.

4. Let A, B be the fixed points, P *any* point in the locus, and C the middle point of AB.

Then for all positions of P the △^{s} ACP, BCP are identically equal [I. 8], so that CP is always perp. to AB.

Hence CP in all its positions lies in the plane through C perp. to AB. [XI. 5.]

Conversely all points in this plane may be shewn to be equidistant from A and B: ∴ the plane through C perp. to AB is the required locus.

5. By the last Ex., the locus of points equidistant from two fixed points A, B is the plane which bisects AB at rt. angles. Hence the point at which the given st. line intersects this plane is that required. The method fails when the given line lies in the above mentioned plane, or is par^l. to it.

6. Let the st. line XY be par^l. to the plane AD, and let any plane BC passing through XY have AB as its common section with the plane AD. Then XY shall be par^l. to AB.

For if not, XY must meet AB at some point Z; but every point in AB is in the plane AD; ∴ XY meets the plane AD at Z; which is impossible, for XY is par^l. to AD.

Page 433.

1. See Def. 4, p. 385.

Let AP, AQ be equal st. lines drawn from A to the plane XY.

Draw AB perp. to the plane XY [XI. 11], and join BP, BQ; then shall the ∠^{s} APB, AQB be equal.

This follows because the △^{s} APB, AQB are identically equal. [Ex. 12, p. 99.]

2. Let A be the given point, BC the given st. line; and let BE be any plane through BC.

From A draw AD perp. to the line BC, and AP perp. to the plane BE [XI. 11]. Then AP, PD, AD lie in a fixed plane through D perp. to BC [XI. 11]. And the ∠ APD is a rt. angle. Therefore the locus of P is a circle on diam^r. AD.

3. Through F draw FH par[l] to BC. [See fig. to XI. 11.]

Then since FH is par[l] to BC, and FDC is a rt. angle [hyp.], ∴ HFD is a rt. angle. And HFA is a rt. angle, for AF is perp. to the plane in which FH is drawn.

Hence FH, being perp. to FA, FD, is perp. to the plane of FDA [XI. 4]; so that BC, being par[l]. to FH, is also perp. to this plane. ∴ BC is perp. to AD in this plane.

Page 439.

1. Take the fig. of Def. 6, p. 412.

Let the dihedral angle PQR between the planes CD, EB be a rt. angle, and let AB be the common section of these planes.

Then PQ is perp. to AB [Def. 7, p. 412 note], and to QR [hyp.]; ∴ PQ is perp. to the plane CD.

And EB is a plane through PQ, ∴ the plane EB is also perp. to the plane CD [XI. 18].

Page 444.

1. Let PA, PB be equal st. lines drawn from the point P to the plane XY. Draw PO perp. to the plane [XI. 11]. Then OA, OB are the projections of PA, PB [Def. 3, p. 410].

In the right-angled △s POA, POB, we have

PA = PB, and PO is common,

∴ OA = OB. [Ex. 12, p. 99.]

2. Let X be any point in SP.

Then in the △s XSA, XSB, XSC,

XS is common, and SA = SB = SC [hyp.].

Also the ∠s XSA, XSB, XSC are equal, being rt. ∠s ;

[Def. 1, p. 410],

∴ XA = XB = XC. [I. 4.]

3. Let A, B, C be the three points; then the lines AB, BC, CA are in one plane [XI. 2].

Find S the centre of the ⊙ circumscribed about the △ABC, and draw SP perp. to the plane of the △ ABC [XI. 12]. Then it may be shewn, as in Ex. 2, that every point in SP produced both ways is equidistant from A, B, and C.

4. Place the rod successively in three positions, so that one of its extremities may be at the given point P and the other in the plane, thus determining three points A, B, C in the plane.

Find S the centre of the ⊙ circumscribed about the △ABC. Then shall S be the foot of the perp. required.

For by Ex. 1, if O is the foot of the perp. from P on the plane, then OA = OB = OC. But there is only one point in the given plane equidistant from A, B, C, namely, the centre of the circumscribed circle. Hence S is the foot of the required perp.

5. Let OA, OB, OC be the three st. lines.

From O cut off along these lines three equal parts OP, OQ, OR; and from O draw OS perp. to the plane PQR [XI. 11].

Then the rt.-angled △s OSP, OSQ, OSR may be shewn identically equal [Ex. 12, p. 99].

∴ ∠ SOA = ∠ SOB = ∠ SOC.

6. Let ABCD be the gauche quadrilateral, and X, Y, Z, V the middle points of the sides AB, BC, CD, DA.

Then ABC, ADC are plane triangles, ∴ XY and VZ are both par[l]. to the common base AC [Ex. 2, p. 104], and are therefore par[l]. to one another [XI. 9].

Similarly it may be shewn by joining BD that XV and YZ are par[l]. Also YZ and VX are in the same plane as XY, ZV [XI. 7].

∴ the figure XYZV is a parallelogram.

7. Through B draw BF par[l]. to AC. Then BF must be in the same plane as AB, AC; and since BAC is a rt. ∠, ∴ FBA is a rt. ∠.

Again, since DB is perp. to the plane of AB, AC, and BF meets it in that plane, ∴ FBD is a rt. ∠.

Hence FB, being perp. to BA and BD, is perp. to the plane of the △ABD [XI. 4]. And since AC is par[l]. to BF, ∴ AC is also perp. to the plane of the △ABD [XI. 8]:

∴ AC is also perp. to AD which meets it in that plane.

8. Let XZ and YV be the two given planes intersecting in the st. line XY; and let these two planes be cut by the first of two par[l]. planes in AP, AQ and by the second in *ap*, *aq*. Then shall the ∠ PAQ = the ∠ *paq*.

Because the par[l]. planes PAQ, *paq* are cut by the plane XZ, ∴ AP and *ap* are par[l]. [XI. 16].

Similarly AQ, *aq* are par[l].

∴ the ∠ PAQ = the ∠ *paq*. [XI. 10.]

9. Let XY be the given plane, and AB the given st. line par[l]. to it. Let the plane AB*ba* pass through AB and cut the given plane XY in the st. line *ab*: then shall *ab* be par[l]. to AB.

For if not, AB and *ab* will meet if produced, since they are in the same plane AB*ba*; but *ab* lies wholly in the plane XY; ∴ AB will meet the plane XY; which is impossible, for AB is given par[l]. to the plane.

Thus AB and *ab*, being in the same plane and not intersecting, are par[l].

10. Let the two planes AY, CY pass one through each of the par[l]. lines AB, CD, and let XY be their common section. Then shall XY be par[l]. to AB and CD.

For if XY be not par[l]. to CD, these lines must intersect at Z, since they are in the same plane.

But XY is in the same plane ABYX; hence Z is in the plane ABYX and also in the plane ABDC; ∴ Z is in AB, their common section. That is, AB and CD intersect at Z; which is impossible, since they are par[l].

Hence XY and CD not intersecting, and being in the same plane, are par[l].: ∴ XY is also par[l]. to AB [XI. 9].

11. Let ABYX, CDYX be two planes, having XY as their common section; and let PQ be a st. line par[l]. to both planes: then PQ shall be par[l]. to XY.

Through PQ take a plane, cutting the plane ABYX in *ab*, and the plane CDYX in *cd*; then *ab* and *cd* are each par[l]. to PQ, and therefore par[l] to one another [Ex. 9, p. 444].

Hence, by Ex. 10, XY is par[l]. to *ab* and *cd*, and therefore to PQ [XI. 9].

12. Let AB, CD be the two st. lines, and P the given point. Take the planes containing AB and P, and CD and P; and let XY be their common section. Then XY shall be the line required.

For since P is a point in each plane, ∴ P lies in XY. And since XY is in a plane with AB, and also in a plane with CD, it intersects both of these lines.

13. Let X, Y, Z be the middle points of AB, BC, CD. Then AC is par[l]. to XY, a line drawn in the plane XYZ; ∴ AC is par[l]. to the plane XYZ; for if AC meet the plane XYZ at some point P, then P would be both in the plane AXYC and in the plane XYZ; that is, P would be in the common section XY, which is impossible, since AC and XY are par[l].

Similarly BD is par[l]. to the plane XYZ.

14. Through E draw EF par[l]. to AB. Then EF is perp. to the plane XY [XI. 8]; hence FEC is a right angle. But AEC is also a rt. angle: ∴ CE is perp. to the plane of EF, EA [XI. 4]. Now EF, EA, AB, EB are in the same plane [XI. 7]; ∴ CE is perp. to EB.

15. Let XYE, XYF be the two planes, having XY as their common section; and let BP, BQ be the common sections of these two planes with the plane of AP, AQ.

Then since AP is perp. to the plane XE, ∴ the plane of AP, AQ is also perp. to the plane XE [XI. 18].

Similarly the plane of AP, AQ is perp. to the plane XF.

Hence the plane of AP, AQ being perp. to the planes XE, XF, is perp. to XY their common section [XI. 19].

16. Let XYE, XYF be the two planes, having the common section XY: and let A be a point in the plane XYE.

Then since AQ is perp. to the plane XF, ∴ the plane APQ is perp. to the plane XF. [XI. 18.]

And since AP is perp. to the plane XE, ∴ the plane APQ is perp. to the plane XE.

∴ the plane APQ, being perp. to the planes XE, XF, is also perp. to XY their common section.

∴ XY is perp. to PQ, a st. line which meets it in the plane APQ.

17. Join AC, BD.

Then the six angles of the two $\triangle^s$ ABC, ADC, namely the $\angle^s$ ABC, ADC, BAC, DAC, BCA, DCA are together equal to *four* rt. angles. [I. 33.]

But the two $\angle^s$ BAC, DAC at the solid angle A are greater than the third $\angle$ BAD. [XI. 20.]

Similarly the two $\angle^s$ BCA, DCA are greater than the $\angle$ BCD.

Hence the four $\angle^s$ ABC, ADC, BAD, BCD are together less than four rt. angles.

18. (i) $\angle$ AOX + $\angle$ BOX greater than $\angle$ AOB [XI. 20]
$\angle$ BOX + $\angle$ COX greater than $\angle$ BOC
$\angle$ COX + $\angle$ AOX greater than $\angle$ COA.

Hence, by addition, twice the sum of the $\angle^s$ AOX, BOX, COX is greater than the sum of the $\angle^s$ AOB, BOC, COA.

(ii) Let OY be the common section of the planes AOB, COX.

Then $\angle$ COB + $\angle$ BOY greater than $\angle$ COY [XI. 20]; to each add $\angle$ YOA.

Then $\angle$ COB + $\angle$ BOA greater than $\angle$ COY + $\angle$ YOA.

But $\angle$ YOA + $\angle$ YOX greater than $\angle$ AOX [XI. 20]; to each add $\angle$ COX.

Then $\angle$ COY + $\angle$ YOA greater than $\angle$ COX + $\angle$ AOX.

A fortiori $\angle$ COB + $\angle$ BOA greater than $\angle$ COX + $\angle$ AOX.

(iii) It has been proved that
$\angle$ AOX + $\angle$ COX less than $\angle$ AOB + $\angle$ BOC;
similarly $\angle$ BOX + $\angle$ AOX less than $\angle$ BOC + $\angle$ COA;
and $\angle$ COX + $\angle$ BOX less than $\angle$ COA + $\angle$ AOB.

Hence, by addition, the sum of the $\angle^s$ AOX, BOX, COX is less than the sum of the $\angle^s$ AOB, BOC, COA.

19. Cf. Ex. 8, p. 102.

In the plane COX and on the side remote from C make the $\angle$ C′OX equal to the $\angle$ COX; and in OC, OC′ take c, c' so that Oc = Oc': then cc' will be bisected perpendicularly by OX at x. Through x in the plane AOB draw axb perp. to Ox meeting OA, OB in a, b. Join ac, bc'.

Then from the $\triangle^s$ cxa, $c'xb$, we have $ac = bc'$. [I. 4.]

Hence from the $\triangle^s$ aOc, bOc', we have $\angle aOc = \angle bOc'$. [I. 8.]

Now $\angle cOc'$ is less than the sum of $\angle^s$ bOc', bOc;

That is, twice $\angle$ COX is less than the sum of $\angle^s$ COA, COB.

20. Let ABC be the $\triangle$ rt. angled at C, O the middle point of AB, and P a point not in the plane of the $\triangle$, such that

$$PA = PB = PC.$$

Then PO shall be perp. to the plane of ABC. Join OC.

Then since ACB is a rt. angle, OA = OB = OC [III. 31].

Hence from the identically equal $\triangle^s$ POA, POB, POC,

$$\angle POA = \angle POB = \angle POC. \quad [I. 8.]$$

But $\angle^s$ POA, POB, being adjacent $\angle^s$ and in the same plane, are rt. angles; $\therefore$ POC is also a rt. angle;

$\therefore$ PO is perp. to the plane ABC. [XI. 4.]

21. Let AB be a st. line drawn from the point A in the plane XY.

Draw BC perp. to the plane, and join AC. Then AC is the projection of AB on the plane.

Let AD be any other line drawn from A in the plane XY.

Then $\angle$ BAC shall be less than $\angle$ BAD.

Make AD equal to AC, and join BD, DC.

Then from the rt.-angled $\triangle$ BCD, BD is greater than BC.

And in the $\triangle^s$ BAC, BAD, we have BA, AC equal to BA, AD respectively, but base BC less than base BD;

$\therefore$ $\angle$ BAC less than $\angle$ BAD. [I. 25.]

22. Let A, B be the points, and XY the plane.

Draw AF perp. to the plane, and produce it to E making FE equal to AF. Join EB cutting the plane in P. Join AP.

Then AP + PB shall be a minimum.

For take any point R in the plane XY, and join AR, RB.

If R is in FP (or FP produced) then AP + PB is less than AR + RB. [Ex. 3, p. 261.] If not draw RQ perp. to FP and join AQ, QB.

Then it may be shewn AP, PB and AQ, QB lie in a plane perp. to XY, and that RQ is perp. to the plane AQB.

Hence AR is greater than AQ, and RB greater than QB.

So that AP + PB is less than AQ + QB [Ex. 3, p. 261];

and AQ + QB less than AR + RB.

23. Let XYE and XYF be two planes having XY as their common section; and let PA, PB be drawn from a point P in the plane XYE so as to be equally inclined to the plane XYF.

From P draw PQ perp. to the plane XYF, and join AQ, BQ.

Then the ∠ PAQ = the ∠ PBQ. [Def. 4, p. 411.]

Hence the △s PAQ, PBQ are identically equal [I. 26];

∴ AP = BP;

∴ the ∠ PAB = the ∠ PBA. [I. 5.]

24. Since PA is perp. to PB, PC, ∴ PA is perp. to the plane BPC [XI. 4]; and PX is drawn perp. to BC in that plane; hence it may be proved that AX is perp. to BC. [Ex. 3, p. 433.]

Similarly BY and CZ are respectively perp. to CA, AB.

∴ XYZ is the pedal △ of the △ ABC.

25. Produce AO, BO, CO to meet BC, CA, AB respectively at X, Y, Z.

Then because AP is perp. to PB, PC, ∴ AP is perp. to the plane PBC.

Hence the plane APXO, which passes through AP, is perp. to the plane PBC. [XI. 18.]

Similarly the plane APXO, which also passes through PO, is perp. to the plane ABC;

∴ BC, the common section of the planes PBC, ABC, is perp. to the plane APXO [XI. 19];

∴ AX is perp. to BC.

Similarly BY, CZ are respectively perp. to CA, AB;

∴ O is the orthocentre of the △ ABC.

MISCELLANEOUS EXERCISES ON SOLID GEOMETRY.
Pages 454—456.

1. Let *ab*, *cd* be the projections of two par[l]. st. lines AB, CD on any plane XY. Then, because A*a*, C*c* are both perp. to plane XY, ∴ A*a* is par[l]. to C*c* [XI. 6]. And AB is par[l]. to CD, ∴ plane BA*a* is par[l]. to plane DC*c* [XI. 15]. But these planes intersect XY in *ab* and *cd* respectively; ∴ *ab* is par[l]. to *cd* [XI. 16].

2. Draw AE par[l] to *ab*. AE will be in plane A*ab*B and will cut B*b* in E. Similarly CF, drawn par[l]. to *cd*, will cut D*d* in F. Because AE and CF are par[l]. respectively to *ab* and *cd* which are par[l]. to one another, ∴ AE is par[l]. to CF [XI. 9]. And because the sides of △ABE are respectively par[l]. to the sides of △CDF, ∴ the ∠[s] of △ABE are equal respectively to the ∠[s] of △CDF [XI. 10]. ∴ AB : CD = AE : CF = *ab* : *cd*.

3. Let AB, CD be the two given st. lines. Through E any pt. in AB, draw EF par[l]. to CD: and through H any pt. in CD, draw HG par[l]. to AB. Then the plane containing AB, EF is par[l]. to the plane containing CD, HG [XI. 15].

4. Let AB, CD be the two given st. lines. As in the last Ex., draw through AB, CD two par[l]. planes. Then it follows from XI. 16 that the projections of AB, CD on *any plane perpendicular to the two par[l]. planes* will be par[l].

5. In the fig. of p. 447, let AB, CD be the given non-intersecting st. lines, having directions at rt. angles to one another; and let HE be the line of constant length. Required the locus of M the middle point of HE.

Draw PQ perp. to AB, CD [Ex. 2, p. 447], and let XY be the plane through AB par[l]. to CD. Draw HK perp. to the plane XY. Join QK, KE; and let the plane through M par[l]. to XY cut PQ, HK at O, S. Then O, S are the middle points of PQ, HK [XI. 17]. Join OM, OS, SM; and draw MN perp. to the plane XY, meeting KE at N. Join QN. Then N is the middle point of KE.

Now in the rt. angled △ HKE, since HE and HK are constant,

∴ KE is constant. [I. 47.]

And in the rt. angled △ KQE, since the hyp. KE is constant, and N is its middle point,

∴ QN = one half of KE = constant. [III. 31.]

But $OM = QN$;

$\therefore$ the locus of M is a $\odot$, of which O is the centre, lying in a plane par[l]. to AB, CD and midway between them.

6. Let O be the angular point.

Then from the rt. angled $\triangle^s$ AOB, AOC, it follows that BO, OC are less than BA, AC respectively. [I. 18.]

But $BC^2 = BO^2 + OC^2$. [I. 47.]

$\therefore BC^2$ is less than $BA^2 + AC^2$;

$\therefore$ the $\angle$ BAC is acute. [Ex. 43, p. 122.]

7. Since the opp. faces of a parallelepiped are parallel, $\therefore$ their common sections with a third plane are parallel [XI. 16].

8. Let dA, dB, dC be three edges terminating in d, and let a, b, c, D be the vertices diagonally opposite to A, B, C, d respectively. Join dD, dc. Then, because each of the planes DcBa and DcAb are perp. to the plane dBcA, $\therefore$ their common section Dc is perp. to the plane dBcA [XI. 19]. $\therefore$ Dc is perp. to dc which meets it in that plane. $\therefore dD^2 = cD^2 + dc^2$. Again, because the planes BdCa, BdAc are each perp. to the plane BcDa, $\therefore$ their common section Bd is perp. to the plane BcDa. $\therefore$ Bd is perp. to Bc, which meets in that plane. $\therefore dc^2 = Bc^2 + dB^2$.

$$\therefore dD^2 = cD^2 + Bc^2 + dB^2 = dC^2 + dA^2 + dB^2,$$

since the faces are parallelograms.

9. Since the edges of a cube are equal, $\therefore$ (diagonal)^2 = three times (edge)^2. [Ex. 8.]

10. See fig. p. 422. In par[m]. ACA′C′,

$$A'A^2 + C'C^2 = 2AC^2 + 2A'C^2 \text{ [Ex. 25, p. 161]};$$

and in par[m]. BDB′D′, $B'B^2 + D'D^2 = 2BD^2 + 2B'D^2$.

$\therefore A'A^2 + B'B^2 + C'C^2 + D'D^2 = 2AC^2 + 2A'C^2 + 2BD^2 + 2B'D^2$. But in par[m]. ABCD, $AC^2 + BD^2 = 2AB^2 + 2BC^2$. $\therefore$ sum of squares on diagonals of par[d] $= 4AB^2 + 4BC^2 + 2A'C^2 + 2B'D^2 = 4AB^2 + 4BC^2 + 4A'C^2$ = sum of squares on twelve edges.

11. Let AP be perp. to base BCD of a regular tetrahedron ABCD. Join BP, CP, DP, and produce them to meet the sides of the base in X, Y, Z.

Then from the rt. angled $\triangle^s$ APB, APC, APD, we have

PB = PC = PD. [Ex. 12, p. 99.]

And from the $\triangle^s$ PBC, PBD, the $\angle$ PBC = the $\angle$ PBD. [I. 8.]

Lastly from the $\triangle^s$ XBC, XBD, we have XC = XD. [I. 4.]

Hence BX is a median of the base: similarly CY, DZ are medians, and P divides each of them in the ratio 2 : 1.

[Ex. 4, p. 113.]

12. Let AP, BQ be perp[s]. from the vertices A, B upon the faces BCD, ACD respectively. Then AQ, BP meet at E, the middle pt. of CD [Ex. 11]. Draw QR par[l]. to AP. This will cut BE, because the par[ls]. AP, QR are in the same plane ABE. And because AP is perp. to plane BCD, so also is QR. ∴ AP : QR = AE : QE = 3 : 1. [Ex. 11.]

13. With the fig. of last Ex., $AE^2 = BE^2 = BC^2 - CE^2 = 3CE^2$.

But $BE = 3PE$, ∴ $BE^2 = 9PE^2$, ∴ $CE^2 = 3PE^2$.

Again, $3AP^2 = 3(AE^2 - PE^2) = 9CE^2 - CE^2 = 8CE^2 = 2a^2$.

14. Let ABCD be the given tetrahedron. Bisect AB in E, CD in E′, AD in F, and BC in F′. Then EF, E′F′ are both par[l]. to BD, ∴ EF is par[l]. to E′F′; and EF′, E′F are both par[l]. to AC, ∴ EF′ is par[l]. to E′F [XI. 9]. ∴ EFE′F′ is a par[m]. ∴ FF′ bisects EE′. Similarly if G, G′ are the middle pts. of AC, BD, GG′ also bisects EE′. ∴ EE′, FF′, GG′ intersect one another at the middle pt. of each.

15. In the tetrahedron ABCD let a plane par[l]. to AC and BD cut the edges AB, BC, CD, DA in the pts. E, F, G, H respectively. Then, because BD is par[l]. to the plane EFGH, ∴ BD is par[l]. to EH, the common section of EFGH with the plane ABD through BD [Ex. 9, p. 444]. Similarly FG is par[l]. to BD. ∴ FG, EH are par[l]. to one another. Similarly EF, HG are par[l]. to one another.

16. Let E, F be the middle pts. of AB, CD, opp. edges of a regular tetrahedron ABCD. Then the $\triangle^s$ CED, AFB being isosceles, EF is perp. to CD and to AB. ∴ EF is the shortest distance between AB and CD. [Ex. 2, p. 447.] Now

$$EF^2 = CE^2 - CF^2 = BC^2 - BE^2 - CF^2 = 2CF^2.$$

$$\therefore 4EF^2 = 8CF^2.$$

But sq. on diagonal of sq. on $CD = 2CD^2 = 8CF^2$. ∴ EF = half diagonal of sq. on edge CD.

17. Let AB be at rt. $\angle^s$ to CD, and AD at rt. $\angle^s$ to BC. Draw BL, CM, DN perp^s. on CD, DB, BC to cut in a the orthocentre of BCD. Then, because CD is at rt. $\angle^s$ to AB and BL, $\therefore$ it is at rt. $\angle^s$ to the plane ABL, and $\therefore$ at rt. $\angle^s$ to Aa in this plane. Similarly BC is at rt. $\angle^s$ to Aa. $\therefore$ Aa is perp. to plane BCD, and $\therefore$ perp. to BD in that plane. $\therefore$ BD is perp. to Aa and aM. $\therefore$ BD is perp. to plane AaM, and $\therefore$ perp. to AC in that plane.

18. By last example, the perp^s. Aa, Cc upon the opp. faces cut those faces in their orthocentres. And the perp^s. upon any edge such as BD from the extremities of the opp. edge AC meet BD in the same pt. M.

(1) Let Aa, Cc cut in X. Then, since a is on CM, and c is on AM, $\therefore$ X is the orthocentre of $\triangle$ACM. $\therefore$ X is the pt. where MM′, the perp. from M upon AC, cuts Aa and Cc. But BD is perp. to plane ACM, $\therefore$ MM′ in this plane is perp. to BD and to AC. $\therefore$ Aa, Cc and the shortest distance between AC and BD cut in the pt. X.

(2) Join BX, and produce it to meet the plane ACD at b. Then, because CD is perp. to AB and Ba, CD is perp. to plane Aba. $\therefore$ plane ACD is perp. to plane Aba, and similarly it is perp. to plane Cbc, $\therefore$ it is perp. to BX the common section of Aba and Cbc. Hence the perp^s. from B on ACD and from D on ABC cut at X. $\therefore$ all the perp^s. are concurrent with one another and with the shortest distance between AC and BD, and therefore with the shortest distances between AB and CD, and between AD and BC.

19. By the last examples,

$$AB^2 = AM^2 + BM^2 \text{ and } CD^2 = CM^2 + DM^2,$$

and

$$BC^2 = BM^2 + CM^2 \text{ and } AD^2 = AM^2 + DM^2.$$

$$\therefore AB^2 + CD^2 = BC^2 + AD^2.$$

20. In the tetrahedron ABCD let P, Q, R be the middle pts. of AB, AC, AD; and L, M, N the middle pts. of CD, DB, BC. Join PL, PC, PD. Then

$$2(DA^2 + DB^2) = 4DP^2 + 4AP^2 \text{ [Ex. 24, p. 161.]}$$
$$= 4DP^2 + AB^2.$$

Similarly $2(CA^2 + CB^2) = 4CP^2 + AB^2$

by addition, $DA^2 + DB^2 + CA^2 + CB^2 = 2DP^2 + 2CP^2 + AB^2$

$= 4PL^2 + CD^2 + AB^2.$

[Ex. 24, p. 161.]

Adding the three similar equations,

$$AB^2 + AC^2 + AD^2 + DB^2 + BC^2 + CD^2 = 4PL^2 + 4QM^2 + 4RN^2.$$

21. Let the plane ACE, which bisects the ∠ between the planes ACB, ACD cut BD in E. Draw DN perp. to plane ACE, and DP, DQ perp. respectively to AC, CE in that plane. Then NP and NQ are perp. respectively to AC and CE [Ex. 14, p. 444]. ∴ the ∠ˢ NPD, NQD are respectively the inclinations of the plane ACE to the planes ACD and BCD. If now B*n*, B*p*, B*q* are drawn perp. respectively to the plane ACE and to the st. lines AC, CE, then the ∠ˢ *np*B, *nq*B are respectively the inclinations of the plane ACE to the planes ACB and BCD. ∴ ∠ *np*B = ∠ NPD and ∠ *nq*B = ∠ NQD. ∴ by similar △ˢ,

$$BE : DE = Bq : DQ = Bn : DN = Bp : DP = \triangle ACB : \triangle ACD.$$

22. If OA, OB, OC are mutually at rt. ∠ˢ, and the △ ABC is equilateral, it may be proved that

$$OA = OB = OC.$$

Take P any point within the △ ABC; and draw PL, PM, PN perp. respectively to the planes OBC, OCA, OAB.

Through P take a plane *obc* par^l. to OBC, and therefore perp. to OA. Then PM and PN lie in the plane *obc*.

Now $PL + PM + PN = PL + oN + Nb$

$= Oo + ob$

$= Oo + oA = OA.$

23. Let ABCD... be the base, and *abcd*... the top of the prism. Let two par^l. planes cut A*a*, B*b*, C*c*, D*d*..., in H, K, L, M... and *h*, *k*, *l*, *m*....

Then HK is par^l. to *hk* [XI. 16], and H*h* is par^l. to K*k* [XI., *Def.* 14.],

∴ HK is par^l. and equal to *hk*.

Similarly KL is par^l. and equal to kl : and so on.

$$\therefore \text{ the } \angle \text{HKL} = \text{the } \angle hkl. \quad [\text{XI. } 10.]$$

In this way it may be shewn that the two polygons have their sides and angles severally equal, ∴ the polygons are equal in all respects.

24. Draw AX perp. to BC, and join OX.

Then OX is also perp. to BC. [Ex. 14, p. 444.]

Now $$\text{AX}^2 . \text{BC}^2 = (\text{AO}^2 + \text{OX}^2)\, \text{BC}^2 \quad [\text{I. } 47.]$$
$$= \text{AO}^2 . \text{BC}^2 + \text{OX}^2 . \text{BC}^2$$
$$= a^2(b^2 + c^2) + b^2c^2 \qquad [\text{Ex. 2, p. } 357.]$$
$$= a^2b^2 + b^2c^2 + c^2a^2.$$

$$\therefore \triangle \text{ABC} = \tfrac{1}{2}\text{AX} . \text{BC} = \tfrac{1}{2}\sqrt{a^2b^2 + b^2c^2 + c^2a^2}.$$

25. See the fig. on p. 45[illegible].

Let P and Q be opp. vertices of the octahedron, and A, B, C, D the remaining vertices.

Then it may be easily proved that the fig. ABCD is a square,

$$\therefore \text{AC}^2 = \text{AD}^2 + \text{DC}^2 = 2\text{AD}^2.$$

Hence $$\text{AC} = \text{AD}\sqrt{2}.$$

26. Let dA, dB, dC be three conterminous edges of the cube, and D, a, b, c the vertices diametrically opposite to d, A, B, C respectively. Bisect Ab, bC, Ca, aB, Bc, cA in E, f, G, e, F, g respectively. Then the sides of the hexagon EfGeFg are clearly equal. And, if X be the middle pt. of cD, then EX, Xe are respectively par^l. to and double of gc, cF ; ∴ Ee is par^l. to and double of gF. Similarly Ee is par^l. to and double of fG. Hence the pts. E, f, G, e, F, g are *co-planar*, and the hexagon EfGeFg is *regular*. [See Book IV. Prop. 15.]

[Four regular plane hexagons are obtained by bisecting all the edges, except those that meet (1) Aa, (2) Bb, (3) Cc, (4) Dd.]

27. Let O be the centre of the sphere.

Draw OC perp. to the plane of section; and take any point P on the line of section of the plane and sphere.

Then $$CP^2 = OP^2 - OC^2 \quad [\text{I. } 47].$$

And since OP and OC are constant, CP is constant.

Hence all points on the line of section are equidistant from C.
∴ the section is a circle, of which C is the centre.

28. See fig. to p. 423.

Since the tetrahedron is regular, the perp^s. from the vertices meet the opp. faces at their centroids. [Ex. 11.]

Hence the perpendiculars meet at a point G, [p. 449.]
where $$Gg_1 = \tfrac{1}{4} Ag_1.$$

But $$3Ag_1^2 = 2a^2 \text{ (Ex. 13)}. \quad \therefore Ag_1 = a\sqrt{\frac{2}{3}},$$

$$\therefore Gg_1 = \frac{a}{4} \cdot \sqrt{\frac{2}{3}} = \frac{a}{\sqrt{6}}.$$

29. Let XY be the given plane, and AB the given st. line.

On AB as diameter describe a sphere. Then it follows from III. 31 that AB subtends a rt. angle at every point on the sphere.

Hence the required locus consists of the points common to the sphere and the plane, and is therefore a circle. [Ex. 27.]

30. Draw ON perp. to given plane, and in ON take A, so that rect. ON, OA = rect. OP, OQ = given constant. ∴ P, Q, A, N are concyclic. And ∠ PNA is a rt. ∠. ∴ ∠ AQO is a rt. ∠. ∴ locus of Q is a sphere described on OA as diameter.

APPENDIX.

Examples on Pole and Polar.

1. Let A and B be the two given points, and let P be the intersection of their polars. Then by the Reciprocal Property of Pole and Polar, since the polar of A passes through P,

$\therefore$ the polar of P passes through A.

Similarly, since the polar of B passes through P,

$\therefore$ the polar of P passes through B.

Hence the polar of P passes through both A and B; that is, AB is the polar of P.

2. Let P be the intersection of the given st. lines PQ, PR, and let A and B be their poles.

Then since AB passes through A, $\therefore$ its pole lies on PQ the polar of A.

Similarly since AB passes through B, $\therefore$ its pole lies on PR the polar of B.

Hence the pole of AB is at P, the only point common to PQ and PR.

3. The locus must be the polar of the given point A; for by the Reciprocal Property of Pole and Polar, (i) the pole of any st. line through A must lie on the polar of A; and (ii) any point on the polar of A must be the pole of some st. line through A.

4. Let O be the common centre, P the point of contact of any one of the tangents, and Q its pole: then since the tangent is perp. to OP [III. 18], Q must lie on OP (or OP produced), and OP . OQ = the sq. on the radius of the given circle. But this radius is constant, and OP is constant, $\therefore$ OQ is constant. Hence the locus of Q is a concentric circle.

5. Let PQ be a diameter of one of the ⊙s, and let O be the centre, and r the radius of the other. From O draw OT touching the first ⊙, and join OP cutting the first ⊙ at R. Join QR.

Now OR . OP = OT^2 [III. 36]

= r^2, since the circles are orthogonal:

and QRP is a rt. ∠, being in a semicircle.

Hence QR is the polar of P: that is, the polar of P passes through Q.

6. Let P and O be the centres of the two ⊙s which intersect at A, B: and let OP cut AB at Q. Join PA, PB.

Then since the ⊙s are orthogonal, PA and PB touch the ⊙ (O) at A and B: hence OP . OQ = $(\text{radius})^2$ [Ex. 1, page 251].

And OP meets the chord of contact at rt. angles [Ex. 2, p. 196].

∴ AB is the polar of P with regard to the ⊙ (O).

7. Let A and B be the given points, and O the centre of the given ⊙. Then since the polars of A and B are respectively perp. to OA, OB, ∴ one of the ∠s between the polars = the ∠ AOB [Ex. 3, p. 65].

8. Let Q be the point inverse to P with respect to the given ⊙. Draw OY perp. to AB; and through Q draw QX perp. to OP, meeting OY at X.

Then since the ∠s at Q and Y are rt. angles,

∴ the points Q, X, Y, P are concyclic [III. 31].

∴ OX . OY = OP . OQ [III. 36]

= r^2 [Hyp.].

But OY is constant, ∴ OX is constant; that is, X is a fixed point.

And since the ∠ OQX is a rt. ∠ [Constr.], ∴ the locus of Q is a circle on OX as diam. [III. 31].

9. Let Q be the point on OP inverse to P, and r the radius of the ⊙ whose centre is O. Draw OX a diam. of the first ⊙. Join PX, and draw QY perp. to OX.

Then OPX is a rt. ∠, being in a semicircle;

and QYX is a rt. ∠ by construction;

∴ the points Q, Y, X, P are concyclic [III. 31];

$$\therefore OX \cdot OY = OP \cdot OQ = r^2.$$

But since OX is constant, ∴ OY is constant;

hence Y is a fixed point.

Therefore the locus of Q is the st. line perp. to OX through the point inverse to X; that is, the polar of X.

10. Let C and D be the points inverse to A and B respectively, and let AX, BY be the perps. from A and B on the polars of B and A. From A and B draw AM, BN perp. respectively to OB and OA (produced if necessary).

Then $OA \cdot OC = OB \cdot OD = r^2$ [Definition].

And since the ∠s at M and N are rt. ∠s, the points M, B, N, A are concyclic,

$$\therefore OA \cdot ON = OB \cdot OM \text{ [III. 36]}.$$

By subtraction

$$OA \cdot NC = OB \cdot DM.$$

But NC = BY, and DM = AX [I. 34].

$$\therefore OA \cdot BY = OB \cdot AX.$$

Examples on Radical Axis and Co-axial Circles.

1. Let TT′ be a common tangent to the two circles, and let their Radical Axis cut TT′ at P. Then, by Definition, the tangential distances of the point P to the two ⊙s are equal: that is, PT = PT′.

2. Let P be any point on the Radical Axis; then the four tangents drawn from P to the two circles are equal [Def.].

Hence a ⊙ described from centre P with any one of these tangents as radius will pass through all four points of contact.

And since the radii drawn from P to the points of contact are also tangents to the given circles, ∴ the ⊙ whose centre is P cuts the given ⊙s orthogonally [p. 240. Def.].

3. As in the last example, all tangents drawn from O to the three ⊙s are equal, ∴ a circle from centre O with radius OT will pass through all the points of contact. And since the radii of this ⊙ drawn to the points of contact are also tangents to the given ⊙s, ∴ the ⊙ whose centre is O cuts the given ⊙s orthogonally.

4. Let the ⊙s (A), (B), (C) touch one another two and two, and let OT, OT′ be the common tangents of the ⊙s (A), (B) and (A), (C) at their points of contact.

Then since OT and OT′ are tangents to the ⊙ (A),

$$\therefore OT = OT'.$$

That is, tangents drawn from O to the ⊙s (B), (C) are equal:

∴ O is a point on the radical axis of the ⊙s (B), (C).

But the radical axis of two ⊙s which touch one another is clearly the common tangent at their point of contact.

Hence the common tangent to the ⊙s (B), (C) passes also through O.

5. Take the figure of p. 243.

Since the ∠s BEA, BEC are rt. angles, ∴ ⊙s described on AB, BC as diams. pass through E [III. 31];

that is, BE is the common chord of the ⊙s on AB and BC.

Similarly AD and CF are respectively the common chords of the ⊙s on AB, AC and on BC, CA.

Hence O, the point of intersection of the common chords, is the radical centre [p. vi. Cor.]

6. See Solution of Ex. 7, p. 252. Observe that the required point B is the *inverse* of the given point A with regard to the given circle.

7. Since by the last Example all ⊙s which pass through the fixed point A and cut a given ⊙ orthogonally, pass also through a second fixed point B (the inverse of A with regard to the given ⊙), ∴ the locus of their centres is the st. line bisecting AB at rt. angles.

To find this point B, draw *any* radius CT to the given ⊙:

describe a ⊙ to pass through A and touch CT at T [Ex. 28, p. 238].

Let this ⊙ cut CA at B. Then B is the required point; for

$$CA \cdot CB = CT^2 \quad [\text{III. } 36].$$

8. Let C be the centre of the given ⊙, and A, D the given points. Now by Ex. 6 all ⊙s through A cutting the given ⊙ orthogonally must pass through B the inverse point of A with respect to the given ⊙.

Determine B as in the last Example. Then the ⊙ circumscribed about the △ABD is that required.

9. Let P be the centre of any ⊙ which cuts the two given ⊙s orthogonally at T and T′.

Then PT = PT′, being radii.

Also PT and PT′ are tangents to the given ⊙s, since the ⊙s are cut orthogonally.

Hence the locus of P is the radical axis of the two given ⊙s.

10. Let C and C′ be the centres of the given ⊙s, and A the given point.

Then all ⊙s through A cutting the ⊙ (C) orthogonally pass through B the inverse of A with respect to the ⊙ (C);

and all ⊙s through A cutting the ⊙ (C′) orthogonally pass through B′ the inverse of A with respect to the ⊙ (C′).

Determine the points B and B′ as in the solution to Ex. 7.

Then the ⊙ about the $\triangle ABB'$ is that required.

Note that by Ex. 9 the centre of this ⊙ is on the radical axis of the given ⊙s (C) and (C′).

11. Let A, B be the centres of the two given ⊙s; PQ, PR tangents to them from the given point P. Let the Radical Axis cut AB at S.

Draw PM, PN perp. respectively to AB and the Radical Axis; and bisect AB at O.

Then $AP^2 - BP^2 = 2AB \cdot OM$ [Ex. 8, p. 159].

And $AQ^2 - BR^2 = AS^2 - SB^2$ [Ex. 1, p. v.]

$= 2AB \cdot OS$ [Ex. 8, p. 159],

∴, by subtraction,

$$AP^2 - AQ^2 - (BP^2 - BR^2) = 2AB\,(OM - OS),$$

or

$$PQ^2 - PR^2 = 2AB \cdot SM = 2AB \cdot PN.$$

12. Let A, B be the centres of two ⊙s of the system, and let their Radical Axis cut AB at S. From P, *any* point in the Radical Axis, draw tangents PQ, PR to the two ⊙s; then PQ = PR [Hyp.]. From centre P, with radius PQ, describe a ⊙ cutting AB at L, L′. Then L, L′ shall be fixed points for all positions of P.

From S draw tangents ST, ST′ to the two ⊙s.

Then $SL^2 = PL^2 - PS^2$ [I. 47]

$= PQ^2 - PS^2$

$= PA^2 - QA^2 - PS^2$

$= PS^2 + AS^2 - QA^2 - PS^2$

$= AS^2 - AT^2$

$= ST^2$.

But ST is independent of the position of P; ∴ L is a fixed point.

Similarly $SL' = ST' = ST = SL$.

GLASGOW: PRINTED AT THE UNIVERSITY PRESS BY ROBERT MACLEHOSE AND CO. LTD.

Spotted something that isn't quite right?

Let us know and we'll fix it for future reprints. And as a thank you for your diligence, we'll send you a free Sothis book of your choice.

publisher@sothis.press

This page intentionally left blank

Made in the USA
Monee, IL
08 November 2026

68001562R00146